AF358624

Computational Methods for Fatigue and Fracture

Computational Methods for Fatigue and Fracture

Editors

Ricardo Branco
Filippo Berto
Shengchuan Wu

MDPI • Basel • Beijing • Wuhan • Barcelona • Belgrade • Manchester • Tokyo • Cluj • Tianjin

Editors
Ricardo Branco
Department of Mechanical
Engineering, CEMMPRE,
University of Coimbra,
3030-788 Coimbra, Portugal

Filippo Berto
Department of Mechanical
and Industrial Engineering,
Norwegian University of
Science and Technology,
7491 Trondheim, Norway

Shengchuan Wu
State Key Laboratory of
Traction Power, Southwest
Jiaotong University,
Chengdu 610031, China

Editorial Office
MDPI
St. Alban-Anlage 66
4052 Basel, Switzerland

This is a reprint of articles from the Special Issue published online in the open access journal *Metals* (ISSN 2075-4701) (available at: https://www.mdpi.com/journal/metals/special_issues/computational_methods_fatigue_fracture).

For citation purposes, cite each article independently as indicated on the article page online and as indicated below:

LastName, A.A.; LastName, B.B.; LastName, C.C. Article Title. *Journal Name* **Year**, *Volume Number*, Page Range.

ISBN 978-3-0365-5299-6 (Hbk)
ISBN 978-3-0365-5300-9 (PDF)

Contents

About the Editors

Ricardo Branco

Ricardo Branco completed his Ph.D. in Mechanical Engineering at the University of Coimbra. He is currently a Professor at the Department of Mechanical Engineering at the University of Coimbra. His research interests include additive manufacturing, structural integrity, fatigue and fracture, multiaxial fatigue, low-cycle fatigue, and fatigue crack growth.

Filippo Berto

Filippo Berto completed his Ph.D. in Mechanical Engineering at the University of Florence. He is currently the Chair of Fatigue and Fracture at the Department of Mechanical Engineering at the Norwegian University of Science and Technology. His research interests include local approaches for fatigue assessment, the fatigue and fracture of advanced materials, multiaxial fatigue, the structural integrity of additive materials, and the numerical modeling of fatigue crack growth.

Shengchuan Wu

Shengchuan Wu completed his Ph.D. in Mechanical Engineering at the Huazhong University of Science and Technology. He is currently the Vice Director of the Railway Vehicle Institute of the State Key Laboratory of Traction Power at Southwest Jiaotong University. His research interests include the fracture and fatigue of advanced materials and advanced characterization based on synchrotron radiation and neutron sources.

Editorial

Computational Methods for Fatigue and Fracture

Ricardo Branco [1],*, Filippo Berto [2],* and Shengchuan Wu [3],*

[1] CEMMPRE, Department of Mechanical Engineering, University of Coimbra, 3030-788 Coimbra, Portugal
[2] Department of Mechanical and Industrial Engineering, Norwegian University of Science and Technology, 7491 Trondheim, Norway
[3] State Key Laboratory of Traction Power, Southwest Jiaotong University, Chengdu 610031, China
* Correspondence: ricardo.branco@dem.uc.pt (R.B.); filippo.berto@ntnu.no (F.B.); wusc@swjtu.edu.cn (S.W.); Tel.: +351-239790700 (R.B.)

1. Introduction and Scope

Fatigue and fracture are pivotal issues in structural integrity. This area of research has attracted much attention over the last years, especially in regard to the main failure mechanisms, the understanding of which is pivotal to developing more durable and more reliable components. Modern numerical methods have been a strong ally in the improvement of current methodologies to assess the integrity of safety-critical components. Moreover, the permanent tendency to shorten time-to-market periods and to reduce overall costs faced by current industry is an additional ingredient for developing alternative simulation approaches because they allow high efficiency at the lowest possible cost. This Special Issue aims to focus on the new trends in computation methods to address fatigue and fracture problems. In this volume, seven papers addressing different research topics have been collected.

2. Contributions

The finite element method plays a key role in the numerical simulation of fatigue and fracture. Alshoaibi and Fageehi [1] developed an adaptive FEM-based approach for simulating the crack advance and the fatigue life in rectangular cross-section plates subjected to tension and bending. The numerical models considered a spider web mesh centred at the crack tip, and the stress intensity factors were computed using displacement extrapolation methods. Li and Xie [2], also basing their research on the finite element method, developed an algorithm to optimize the tooth surface contact stress in spur gears, considering tooth profile deviations, meshing errors, and lead crowning modifications. The problem was addressed using a three-dimensional model of one of the engaged teeth, by combining an optimized area of high refinement level with a non-refined area connected via multi-point constraint.

The analysis of critical engineering components by combining the finite element method with advanced fatigue methods was another line of research. Concli et al. [3] developed a critical-plane approach in conjunction with three-dimensional FEM models to study the early crack propagation stage in teeth subjected to bending fatigue. The models, created from extruded meshes and capable of accounting for boundary effects, allowed the determination of crack direction and the critical region associated with the crack nucleation. Sánchez et al. [4] determined the load-bearing capacity of tubular beams made of aluminium by applying the theory of critical distances and linear-elastic two-dimensional finite element models. The proposed methodology has been successfully validated for cantilever beams with circumferential U-shaped notches, leading to errors in the predicted load-bearing capacity lower than ±20%.

Computational tools can play an important role in optimization problems. The paper by Khan et al. [5] presents a simulation-based optimization methodology for mold design and the prediction of reliability in mechanical components with minimum level of casting

Citation: Branco, R.; Berto, F.; Wu, S. Computational Methods for Fatigue and Fracture. *Metals* **2022**, *12*, 739. https://doi.org/10.3390/met12050739

Received: 11 April 2022
Accepted: 24 April 2022
Published: 27 April 2022

Publisher's Note: MDPI stays neutral with regard to jurisdictional claims in published maps and institutional affiliations.

defects. The reliability was computed using classical strength–stress models and probability distribution functions. Woo et al. [6] proposed a parametric accelerated life testing approach to improve the fatigue life of mechanical components subjected to impact loading. The concept was tested in a domestic refrigerator hinge kit system, leading to a new design and an extended fatigue life. These promising results make the proposed parametric accelerated life testing approach applicable to metallic parts of other machines, namely cams, gears, crankshafts, and dies, to mention just a few.

In the context of fatigue crack propagation, the extended finite element method (XFEM) allows the alleviation of the shortcomings of the finite element method, namely with regard to the modelling of cracks and material interfaces. These advantages were explored by Fageehi [7], who studied the fatigue crack growth under mixed-mode loading in modified four-point bending beams and cracked plates with three holes. The fatigue assessment was conducted using fracture mechanics, stress-life methods, and strain-life methods. The proposed methodology was capable of simulating the crack paths, calculating the mixed-mode stress intensity factors at the crack front, and estimating the fatigue life for different geometrical configurations.

3. Conclusions and Outlook

The present Special Issue is aimed at collecting original contributions on the new trends in computation methods, which address fatigue and fracture problems. Seven papers were selected to cover a wide variety of current trends and applications. The topics addressed demonstrate that the finite element method remains a powerful technique in this field. However, other advanced tools are immerging, such as the extended finite element method or the meshless methods. In addition, the topics collected demonstrate the richness and the potential of current computational methods to deal with complex fatigue and fracture problems from different perspectives, namely the simulation of crack propagation in gears, the prediction of crack paths in notched plates, the calculation of critical loads in notched components subjected to different loading histories, the improvement of mold design and minimization of casting defects, and the development of advanced mechanical systems subjected impact loading. However, since numerical simulation of fatigue and fracture phenomena involves a myriad of problems, there are still open challenges that need to be addressed in order to translate scientific research into practical and daily life applications. In this context, guested edited by the same team, a new Special Issue entitled "Numerical Methods Applied to Fatigue and Fracture Phenomena" (https://www.mdpi.com/journal/metals/special_issues/numerical_fatigue_fracture) will be launched in *Metals*, to collect and disseminate the future advances in these areas.

Conflicts of Interest: The authors declare no conflict of interest.

References

1. Alshoaibi, A.M.; Fageehi, Y.A. Simulation of Quasi-Static Crack Propagation by Adaptive Finite Element Method. *Metals* **2021**, *11*, 98. [CrossRef]
2. Li, Q.; Xie, L. Analysis and Optimization of Tooth Surface Contact Stress of Gears with Tooth Profile Deviations, Meshing Errors and Lead Crowning Modifications Based on Finite Element Method and Taguchi Method. *Metals* **2020**, *10*, 1370. [CrossRef]
3. Concli, F.; Maccioni, L.; Fraccaroli, L.; Bonaiti, L. Early Crack Propagation in Single Tooth Bending Fatigue: Combination of Finite Element Analysis and Critical-Planes Fatigue Criteria. *Metals* **2021**, *11*, 1871. [CrossRef]
4. Sánchez, M.; Cicero, S.; Arroyo, B.; Álvarez, J.A. Coupling Finite Element Analysis and the Theory of Critical Distances to Estimate Critical Loads in Al6060-T66 Tubular Beams Containing Notches. *Metals* **2020**, *10*, 1395. [CrossRef]
5. Khan, M.A.A.; Sheikh, A.K.; Gasem, Z.M.; Asad, M. Fatigue Life and Reliability of Steel Castings through Integrated Simulations and Experiments. *Metals* **2022**, *12*, 339. [CrossRef]
6. Woo, S.; O'Neal, D.L.; Woldemichael, D.E.; Atnaw, S.M.; Tulu, M.M. Improving the Fatigue of Newly Designed Mechanical System Subjected to Repeated Impact Loading. *Metals* **2021**, *11*, 139. [CrossRef]
7. Fageehi, Y.A. Fatigue Crack Growth Analysis with Extended Finite Element for 3D Linear Elastic Material. *Metals* **2021**, *11*, 397. [CrossRef]

Review

Simulation of Quasi-Static Crack Propagation by Adaptive Finite Element Method

Abdulnaser M. Alshoaibi * and Yahya Ali Fageehi

Mechanical Engineering Department, Jazan University, Jazan 45142, Saudi Arabia; yfageehi@jazanu.edu.sa
* Correspondence: alshoaibi@jazanu.edu.sa or alshoaibi@gmail.com; Tel.: +966-56-347-9523

Abstract: The finite element method (FEM) is a widely used technique in research, including but not restricted to the growth of cracks in engineering applications. However, failure to use fine meshes poses problems in modeling the singular stress field around the crack tip in the singular element region. This work aims at using the original source code program by Visual FORTRAN language to predict the crack propagation and fatigue lifetime using the adaptive dens mesh finite element method. This developed program involves the adaptive mesh generator according to the advancing front method as well as both the pre-processing and post-processing for the crack growth simulation under linear elastic fracture mechanics theory. The stress state at a crack tip is characterized by the stress intensity factor associated with the rate of crack growth. The quarter-point singular elements are constructed around the crack tip to accurately represent the singularity of this region. Under linear elastic fracture mechanics (LEFM) with an assumption in various configurations, the Paris law model was employed to evaluate mixed-mode fatigue life for two specimens under constant amplitude loading. The framework includes a progressive analysis of the stress intensity factors (SIFs), the direction of crack growth, and the estimation of fatigue life. The results of the analysis are consistent with other experimental and numerical studies in the literature for the prediction of the fatigue crack growth trajectories as well as the calculation of stress intensity factors.

Keywords: LEFM; mesh density; mixed mode stress intensity factors; fatigue crack growth; FEM

Citation: Alshoaibi, A.M.; Fageehi, Y.A. Simulation of Quasi-Static Crack Propagation by Adaptive Finite Element Method. *Metals* **2021**, *11*, 98. https://doi.org/10.3390/met11010098

Received: 17 December 2020
Accepted: 3 January 2021
Published: 6 January 2021

Publisher's Note: MDPI stays neutral with regard to jurisdictional claims in published maps and institutional affiliations.

1. Introduction

The finite element method (FEM) is definitely the most common and effective analytical technique for analyzing the behavior of a wide variety of engineering and physical issues. One of the essential uses of FEM is the study of crack propagation. The propagation of the crack reduces components' ability to resist the external load and eventually break the components. Analyzing fatigue crack growth is necessary to ensure the stability of structures subjected to cyclic loading. Cracks begin due to the presence of plastic strain caused by cyclic tension, and they grow due to the tensile stress. However, compressive loads do not lead to fatigue cracks due to the local tensile stress [1]. A variety of software has been developed for general purposes for finite elements, verified and calibrated through the years and now available on request, the most well-known being three-dimensional, such as ANSYS [2], ABAQUS [3], NASTRAN, FRANC3D, and COMSOL. In addition, there are numerous 2D simulation software for crack propagation simulation, e.g., NASGRO, AFGROW, FRANC2D, and FASTRAN. Many researchers have also developed an effective method for estimation of fatigue breakage growth in 2D linear elastic structures with multi-mode loading [4–7]. Determining the accurate stress intensity factor of a cracked structure in LEFM is very crucial in accessing the integrity of the crack, especially if the calculation is carried out using the finite element technique with extremely fine mesh. The propagation of the crack can be simulated at the highest accuracy by increasing the mesh density, as well as estimating the stress intensity factors accurately. In addition, very fine mesh around the crack tip is needed for precise prediction of SIFs using a nodal displacement technique such as the displacement extrapolation technique (DET). The DET requires configuration of special elements in the vicinity of the crack tip, by correctly representing the stress field singularity at the crack tip. These special elements, known as singular elements, need to be constructed in a rosette formation around every crack tip. Very small-size elements can be optimally created around the crack tip with the use of an adaptive mesh refinement scheme. Generating overall fine mesh leads to greater computational time. This procedure was reduced by using the adaptive mesh strategy, which increases the mesh only on the required areas. The adaptive mesh refinement scheme is another method to generate the optimal mesh in a very efficient way. Many studies on mesh refinement problems and related errors in computing SIFs using the FEM were conducted [8,9]. Another study [10] was been performed to clarify the effect of the in-plane and out-of-plane constraints on the ductile fracture with different crack sizes, specimen thicknesses, and span lengths. They concluded that the lower in-plane and out-of-plane constraint levels introduce higher fracture properties. It is more challenging to combine the extreme fine mesh generation with the adaptive scheme and solve the stiffness equation matrix. The benefits sought here are both faster execution time and the ability to process larger problems. In order to simulate 2D cracks under mixed mode loading, the current developed software code is formulated to allow the researcher to estimate the fatigue life and crack trajectory using the automated adaptive mesh finite element [11–15]. This software was created in 2004 and continues to include several features for the simulation of two-dimensional fatigue crack growth under LEFM assumptions [12,16–21]. The use of commercial software for engineers is not appropriate in at least two aspects: First, the basic algorithm that lies behind it is not fully understood, and second, the execution is completely apprehended throughout the programming ability. Commercial software can be used to model crack propagation as well, but such software is very expensive and can hardly get the source code to develop it.

2. Developed Program Framework

The code that was developed is a simulation software to assess the 2D crack propagation process under LEFM conditions. This software predicts the growth of quasi-static crack growth in 2D components using the finite element method, taking into account the mechanical parameters of the fracture. Four essential features for the adaptive mesh finite element (FE) analysis are used for the crack direction simulation, namely, the mesh optimization algorithm, the crack criteria, the criterion of direction, and the methodology of crack propagation. The mesh refining can be controlled by the characteristic scale of each element predicted, based on the current error estimator. An incremental principle with the von Mises yield criterion is applied to this initial model. The solution errors are computed after each load stage is over. The incremental analysis is interrupted when the error exceeds a specified cumulative error at some stage and a new FE plan is generated. The program automatically configures the mesh with a new mesh refinement. After it is generated, the solution variables (displacement, stresses, strains, etc.) are transferred from the old mesh to the new mesh. The analysis is then resumed and progresses until the errors are again higher than the pre-decided amount.

In order to examine the start of the crack growth, the crack growth criterion is employed. The LEFM typically utilizes SIFs as a fracture criterion. Various techniques of estimating the path of a crack are used, such as the maximum circumferential stress theory, theory of maximum energy release, and theory of minimum energy density. At any stage of crack propagation, a FE model is defined. This model is given in the first step as an input for the modeling. The algorithm output is then generated via the models in the subsequent steps. At each stage, as the crack grows, the geometry elements are deleted and reconstructed using an adaptive technique and updated for the next propagation process. Figure 1 demonstrates the simulation procedure used to model quasi-static crack growth. The main steps of this procedure are explained in detail by [11,14].

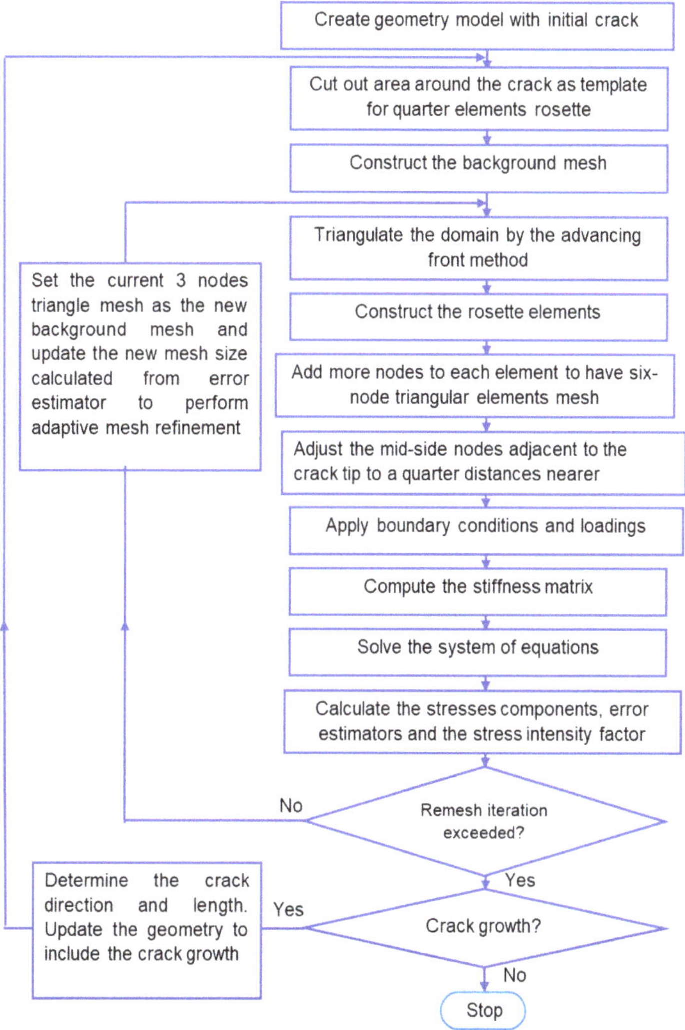

Figure 1. General flow chart of the quasi-static crack growth program.

2.1. Displacement Extrapolation Technique (DET)

The DET is based on the nodal displacement around the crack tip. The construction of quarter-point elements around the crack tip is generally needed for this procedure. Generally, the existence of the quarter-point element is essential in order to correctly represent the linear elastic singularity $(1/\sqrt{r})$ for stresses and strains at the crack tip. The polynomial isoparametrically representative of the singularity is typically obtained by moving the mid-side nodes adjacent to the crack tip to a quarter-length edge closer to the crack tip. Crack tip elements based on this method were separately suggested by [22,23]. In this study, the natural triangle–quarter-point element was selected as the type of crack-tip element and its configuration follows the schematic formation of the rosette around the crack-tip, as seen in Figure 2.

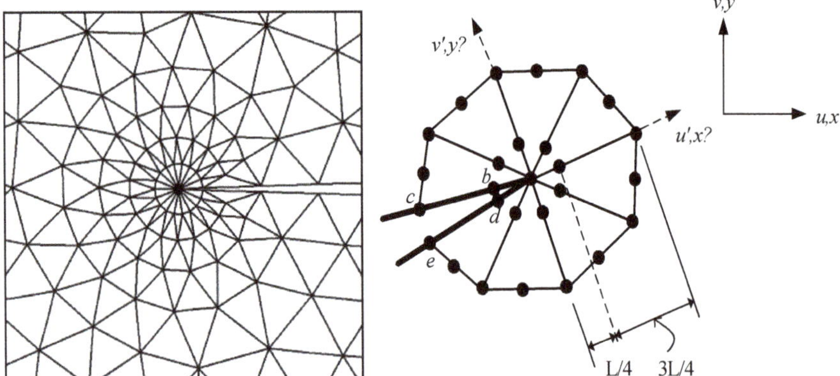

Figure 2. A quarter-point singular element around the tip of the crack.

For the calculation of stress intensity factors, the displacement extrapolation method [24] was used as follows:

$$K_I = \frac{E}{3(1+v)(1+\kappa)} \sqrt{\frac{2\pi}{L}} \left[4(v'_b - v'_d) - \frac{(v'_c - v'_e)}{2} \right] \tag{1}$$

$$K_{II} = \frac{E}{3(1+v)(1+\kappa)} \sqrt{\frac{2\pi}{L}} \left[4(u'_b - u'_d) - \frac{(u'_c - u'_e)}{2} \right] \tag{2}$$

where E is the modulus of elasticity, v is the Poisson's ratio, κ is the elastic parameter defined by

$$\kappa = \begin{cases} 3 - 4v \text{ for plane strain} \\ \frac{(3-v)}{(1+v)} \text{ for plane stress} \end{cases} \tag{3}$$

and L is the quarter-point element length. The u' and v' are the displacement components in the x' and y' directions, respectively. The subscriptions represent their position, as seen in Figure 2.

2.2. Adaptive Mesh Refinement

To minimize expected errors after a finite element solution has been achieved, an adaptive mesh refinement technique is used. The method of adaptive mesh refinement measures the mesh's adequacy and refines the mesh wherever the estimated error is large. Until user-definable error tolerance is reached, the system iterates the mesh refinement and solution. Because the precision of the solution depends on these tolerance limits, it is important for the use of adaptive mesh generators to provide a good understanding of the FEM in an effective manner. The method is referred to as adaptive since at all times the process relies on previous results. The adaptive remeshing method was carried out on the basis of the posteriori stress error standard scheme to achieve the optimum mesh from [16]. The software adopted a frontal solver that is an effective direct solver used to solve a linear equation system. In h-type adaptive mesh refinement, the major point is to obtain the ratio of element normal stress error to the average normal stress error for the entire domain, which is also known as the relative stress norm error, and a new size can be predicted from this ratio for the refinement method. The mesh size is defined in the procedure of each element as:

$$h_e = \sqrt{2A_e} \tag{4}$$

where A_e is the area of the triangle element. The norm stress error for each element is defined by

$$\|e\|_e^2 = \int_{\Omega^e} (\boldsymbol{\sigma} - \boldsymbol{\sigma}^*)^{\mathrm{T}} (\boldsymbol{\sigma} - \boldsymbol{\sigma}^*) d\Omega$$

$$= \int_{\Omega^e} \left(\left\{ \begin{array}{c} \sigma_x \\ \sigma_y \\ \tau_{xy} \\ \sigma_z \end{array} \right\} - \left\{ \begin{array}{c} \sigma_x^* \\ \sigma_y^* \\ \tau_{xy}^* \\ \sigma_z^* \end{array} \right\} \right)^{\mathrm{T}} \left(\left\{ \begin{array}{c} \sigma_x \\ \sigma_y \\ \tau_{xy} \\ \sigma_z \end{array} \right\} - \left\{ \begin{array}{c} \sigma_x^* \\ \sigma_y^* \\ \tau_{xy}^* \\ \sigma_z^* \end{array} \right\} \right) d\Omega \tag{5}$$

whereas the average norm stress error for the whole domain is

$$\|\hat{e}\|^2 = \frac{1}{m} \sum_{e=1}^{m} \int_{\Omega^e} \boldsymbol{\sigma}^{\mathrm{T}} \boldsymbol{\sigma} d\Omega$$

$$= \frac{1}{m} \sum_{e=1}^{m} \int_{\Omega^e} \left\{ \begin{array}{c} \sigma_x \\ \sigma_y \\ \tau_{xy} \\ \sigma_z \end{array} \right\}^{\mathrm{T}} \left\{ \begin{array}{c} \sigma_x \\ \sigma_y \\ \tau_{xy} \\ \sigma_z \end{array} \right\} d\Omega \tag{6}$$

where m indicates the total number of elements in the whole domain and σ^* is the smoothed stress vector. In the finite element treatment the integration with the isoparametric triangular element is converted by the summation of quadratures following the Radau rules [25] as follows:

$$\|e\|_e^2 = \int_{-1}^{1} \int_{-1}^{1} \left(\left\{ \begin{array}{c} \sigma(\xi,\eta)_x \\ \sigma(\xi,\eta)_y \\ \tau(\xi,\eta)_{xy} \\ \sigma(\xi,\eta)_z \end{array} \right\} - \left\{ \begin{array}{c} \sigma(\xi,\eta)_x^* \\ \sigma(\xi,\eta)_y^* \\ \tau(\xi,\eta)_{xy}^* \\ \sigma(\xi,\eta)_z^* \end{array} \right\} \right)^{\mathrm{T}} \left(\left\{ \begin{array}{c} \sigma(\xi,\eta)_x \\ \sigma(\xi,\eta)_y \\ \tau(\xi,\eta)_{xy} \\ \sigma(\xi,\eta)_z \end{array} \right\} - \left\{ \begin{array}{c} \sigma(\xi,\eta)_x^* \\ \sigma(\xi,\eta)_y^* \\ \tau(\xi,\eta)_{xy}^* \\ \sigma(\xi,\eta)_z^* \end{array} \right\} \right) t^e \det J^e d\xi d\eta$$

$$= \sum_{p=1}^{3} \left(\left\{ \begin{array}{c} \sigma(\xi_p,\eta_p)_x \\ \sigma(\xi_p,\eta_p)_y \\ \tau(\xi_p,\eta_p)_{xy} \\ \sigma(\xi_p,\eta_p)_z \end{array} \right\} - \left\{ \begin{array}{c} \sigma(\xi_p,\eta_p)_x^* \\ \sigma(\xi_p,\eta_p)_y^* \\ \tau(\xi_p,\eta_p)_{xy}^* \\ \sigma(\xi_p,\eta_p)_z^* \end{array} \right\} \right)^{\mathrm{T}} \left(\left\{ \begin{array}{c} \sigma(\xi_p,\eta_p)_x \\ \sigma(\xi_p,\eta_p)_y \\ \tau(\xi_p,\eta_p)_{xy} \\ \sigma(\xi_p,\eta_p)_z \end{array} \right\} - \left\{ \begin{array}{c} \sigma(\xi_p,\eta_p)_x^* \\ \sigma(\xi_p,\eta_p)_y^* \\ \tau(\xi_p,\eta_p)_{xy}^* \\ \sigma(\xi_p,\eta_p)_z^* \end{array} \right\} \right) t^e \det J^e W_p \tag{7}$$

where W_P is a weighting factor and is J^e is the Jacobian matrix.
Similarly,

$$\|\hat{e}\|^2 = \frac{1}{m} \sum_{e=1}^{m} \sum_{p=1}^{3} \left(\left\{ \begin{array}{c} \sigma(\xi_p,\eta_p)_x \\ \sigma(\xi_p,\eta_p)_y \\ \tau(\xi_p,\eta_p)_{xy} \\ \sigma(\xi_p,\eta_p)_z \end{array} \right\} \right)^{\mathrm{T}} \left(\left\{ \begin{array}{c} \sigma(\xi_p,\eta_p)_x \\ \sigma(\xi_p,\eta_p)_y \\ \tau(\xi_p,\eta_p)_{xy} \\ \sigma(\xi_p,\eta_p)_z \end{array} \right\} \right) t^e \det J^e W_p \tag{8}$$

where t^e is the element thickness for a plane stress condition and $t^e = 1$ for a plane strain condition. Therefore, the relative stress norm error ζ_e for each item is considerably less than some identified value [26]. Thus,

$$\zeta_e = \frac{\|e\|_e}{\|\hat{e}\|} \leq \zeta \tag{9}$$

And the relative stress error level of the new element is defined as permissible error as

$$\varepsilon_e = \frac{\|e\|_e}{\zeta \|\hat{e}\|} \leq 1 \tag{10}$$

This implies that any element with $\varepsilon_e > 1$ must be optimized and the new mesh size must be predicted. The asymptotic convergence rate criteria are used, which assumed

$$\|e\|_e \propto h_e^p \tag{11}$$

where p is the approximation of the polynomial order. In the analysis, $p = 2$ is used for the approximation of finite elements as a quadratic polynomial. The predicted sizes of the new element are stated as follows:

$$h_N = \frac{1}{\sqrt{\varepsilon_e}} h_e \tag{12}$$

where h_e is the old element size and p is the order of the interpolation shape function.

Convergence of the mesh is dependent of the size of the new element, which defines how many elements in a model are required to ensure that the results of an analysis are not affected by changing the mesh size. System response (stress, deformation) converge with decreasing element size to a repeatable solution. Further refinement of the mesh does not affect results because the model and its results are now independent of the mesh.

The present mesh is known as the new background mesh and the advancing front method is replicated according to the amount of mesh refinements set by the user.

The mesh optimization is used in the final stage of the mesh generation in order to enhance the shape of the elements. The topological structure of the mesh is fixed in the process of mesh smoothing, i.e., the element's nodal connections are not changed, but the inner nodes are repositioned to create triangles with much improved shapes. The most effective computational smoothing algorithm is the well-known Laplacian smoothing [27], which repositions the inner node created by its neighboring nodes at the center of the polygon. The new position of an internal node i is computed as

$$(x_i, y_i) = \frac{1}{N_n} \sum_{j=1}^{N} (x_j, y_j) \tag{13}$$

where N_n is the number of nodes linked to node i. The mesh smoothing process consists of several iterations.

2.3. Crack Growth Analysis

The direction of the crack path under linear elastic conditions must be computed to facilitate crack propagation simulation. The maximum circumferential stress theory states that for isotropic materials under mixed loading mode the crack grows in a direction normal to a maximum tangential tensile stress. The tangential stress is estimated in polar coordinates as

$$\sigma_\theta = \frac{1}{\sqrt{2\pi r}} \cos \frac{\theta}{2} \left[K_I \cos^2 \frac{\theta}{2} - \frac{3}{2} K_{II} \sin \theta \right] \tag{14}$$

The direction normal to the tangential maximum stress can be obtained by resolving $d\sigma_\theta/d\theta = 0$ for θ. The nontrivial solution is determined by

$$K_I \sin \theta + K_{II}(3 \cos \theta - 1) = 0 \tag{15}$$

which can be solved as

$$\theta = \pm \cos^{-1} \left\{ \frac{3K_{II}^2 + K_I \sqrt{K_I^2 + 8K_{II}^2}}{K_I^2 + 9K_{II}^2} \right\} \tag{16}$$

The sign of θ must be opposite the sign of K_{II} to ensure the optimal opening stress associated with the crack direction [28]. Figure 3 illustrated the two possibilities of the crack growth direction.

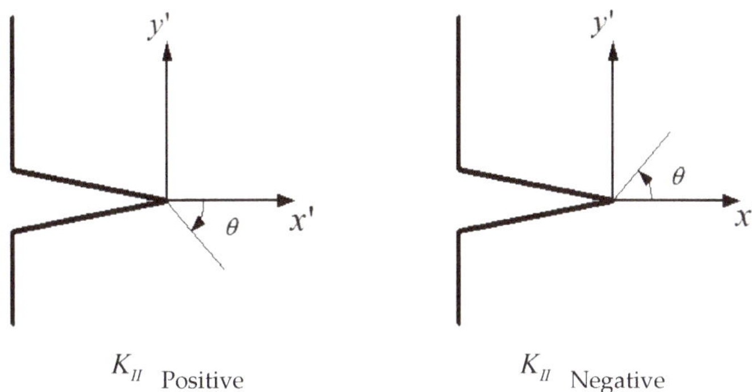

Figure 3. Sign of the crack growth angle.

In the case of fatigue crack growth, the resulting stress intensity range at each crack tip must exceed the stress intensity threshold, specified as

$$\Delta K_{th} = f \Delta \sigma_{th} \sqrt{\pi a} \tag{17}$$

where f is a geometrical and loading function and $\Delta \sigma_{th}$ is the stress range limit. According to Equation (17), the crack is not propagated if $\Delta \sigma < \Delta \sigma_{th}$. This equation was practically modified by using another parameter known as the equivalent stress intensity factor range, ΔK_{Ieq}. Therefore, if $\Delta K_{Ieq} > \Delta K_{th}$, this indicates commencement of fatigue crack growth. This parameter is set to

$$\Delta K_{Ieq} = \Delta K_I \cos^3(\theta/2) - 3\Delta K_{II} \cos^2(\theta/2) \sin(\theta/2) \tag{18}$$

In the modified equation of the Paris law, Tanaka [20] derived an innovative law known as the power law for determining crack growth in response to fatigue with the equivalent stress intensity factor (ΔK_{eq}) as

$$\frac{da}{dN} = C(\Delta K_{eq})^m \tag{19}$$

where a is the length of the crack, N is the number of cycles, C is the Paris constant (mm/cycle), and m is the Paris exponent.

The total number of fatigue lifecycles can be calculated using Equation (19) for an increase in crack length as

$$\int_0^{\Delta a} \frac{da}{C(\Delta K_{eq})^m} = \int_0^{\Delta N} dN = \Delta N \tag{20}$$

3. Numerical Results and Discussion

3.1. Two Internal Non-Colinear Cracks

For this geometry, there were two internal, parallel, non-colinear, and non-angled cracks in a rectangular specimen with dimensions (90 mm/180 mm). The initial crack length was a = 10 mm for both cracks. As seen in Figure 4a, this geometry was subjected to acyclic tension ($\sigma_{max} = 160$ N/mm, $\sigma_{min} = 0$) at the upper end and restricted at the bottom side. The distance between the two tips was 15 mm in the horizontal direction and 5 mm in the vertical direction. The adaptive dense mesh is shown in Figure 4b. The selected material was aluminum, which has the material properties shown in Table 1.

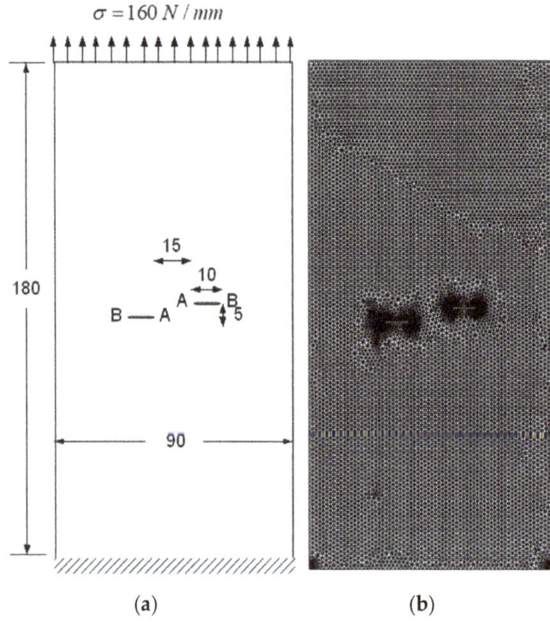

Figure 4. (**a**) Problem statement for two internal non-colinear cracks (all dimensions in mm) and (**b**) adaptive mesh for the specimen.

Table 1. Material properties of aluminum.

Property	Value in Metric Unit
Modulus of elasticity, E	74 GPa
Poisson's ratio, υ	0.3
Fracture toughness, K_{IC}	60 MPa$\sqrt{m}$
Threshold stress intensity factor, K_{th}	4 MPa$\sqrt{m}$
Paris law coefficient, C	2.087136×10^{-13}
Paris law exponent m	3.32

This specimen contained four crack tips, which made it interesting to observe the interaction between cracks and to further explore the performance of the developed software in the simulation of multiple cracks.

The predicted crack growth is shown in Figure 5a, which closely resembled the experimental result of Tu and Cai (1993), as illustrated in Figure 5c. These predicted crack growth trajectories were also in agreement with the numerical results obtained by [29] using the linear smoothed extended finite element method, which was compared to the numerical results reported by [5] using a meshless method with enriched weight functions, as shown in Figure 5b.

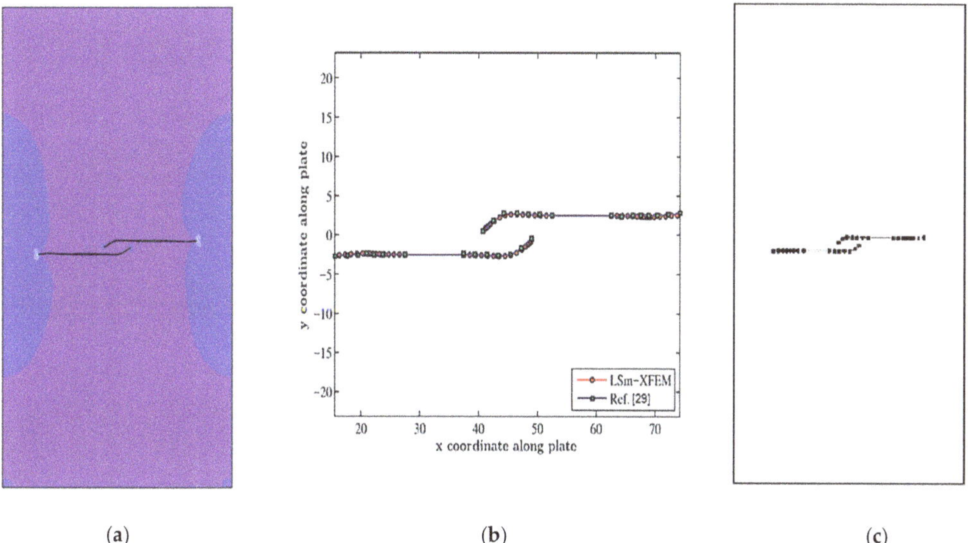

(a)   (b)   (c)

Figure 5. (**a**) Crack propagation simulation for the two internal non-colinear cracks specimen, (**b**) the numerical results of [29], with permission from Elsevier 2019, and (**c**) the experimental results [30].

Figure 6 compares stress intensity factors in tips A and B along the crack length with the result from the meshless finite element [5]. Actually, the crack length values were the cumulative crack increment in tips A and B, starting at 10 mm in each stage, which was the original crack length. Only the upper right slip result was selected in the graph. The figure shows good agreement for the comparison results. The deviation of K_I^A at a crack length of 27 mm above was attributable to the contact with the opposite crack trajectories.

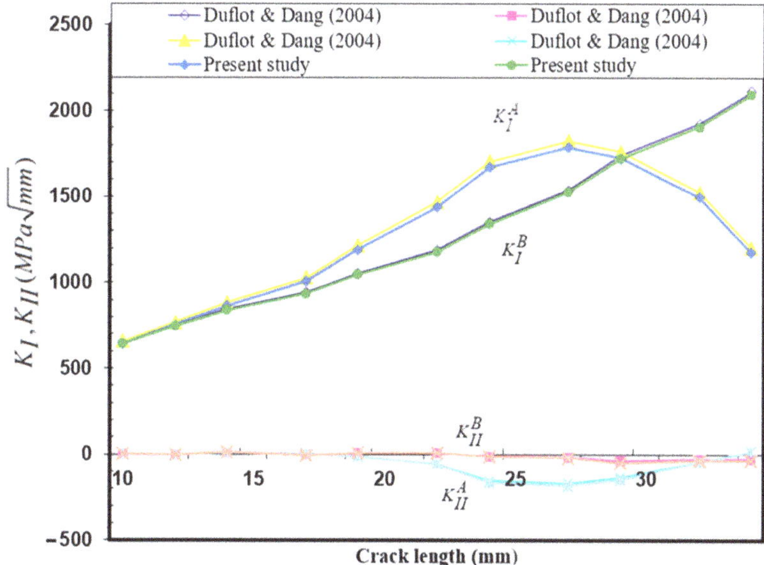

Figure 6. SIFs versus crack length comparison between the present study and the results of [5] for the two internal non-colinear crack specimens.

Both cracks demonstrated in the beginning a pure mode I of approximately the same SIF values. After that, the mode II of the SIF increased at tip A above that of tip B while the second mode of SIFs became negative at A, thus making the crack path curve towards the other break. Eventually the second mode of the SIFs at A tended to decrease as crack tip A moved closer when the first mode at B increased continually. Finally, the equivalent mode I of the SIF at B exceeded the fracture toughness and unstable fracture occurred at crack tip B. The fatigue life of the structure was evaluated as 6840 cycles, which was in good agreement with the results obtained by [5] using a meshless method, as shown in Figure 7, as well as with the numerical results obtained by [29].

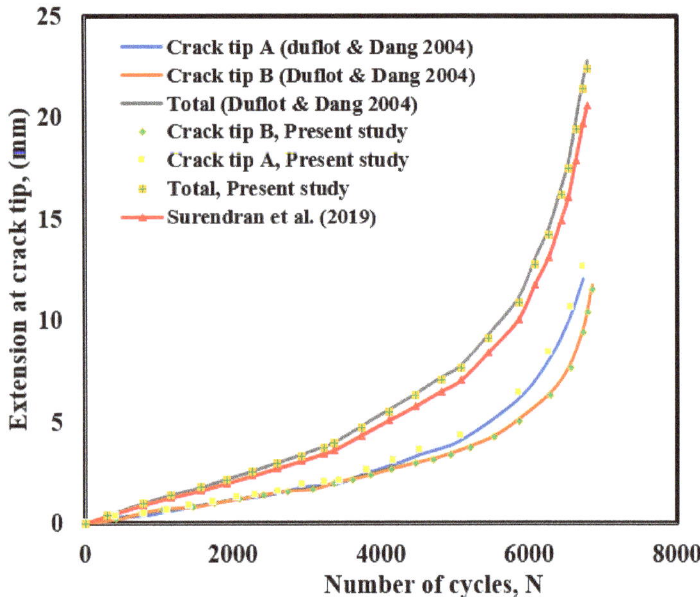

Figure 7. Comparison for the fatigue lifecycles for the two internal non-colinear cracks.

3.2. PMMA Beam Specimen

The PMMA beam geometry offers a benchmark evaluation based on the numerical and experimental work of [31]. The beams were made of polymethyl methacrylate (PMMA), which is a standard material option for crack path investigations as it is relatively homogeneous and exhibits brittle fracture behavior at room temperature. The specimen was under a cyclic point load and acted on the top mid-span position with a value of 4.448 kN. The properties of the materials were taken as modulus of elasticity, $E = 205$ GPa, yield stress $\sigma_y = 516$ MPa, threshold stress intensity factor $\Delta k_{th} = 80$ MPa$\sqrt{\text{mm}}$, $\Delta K_{IC} = 730$ MPa$\sqrt{\text{mm}}$, Paris law coefficient, $C = 1.2 \times 10^{-11}$, Paris law exponent m = 3, and Poisson's ratio $\nu = 0.3$. The thickness of the specimen was 12.7 mm and there were two different configurations depending on the initial crack length (a) and its position (b), as shown in Table 2. The specimen's geometry and the initial adaptive dens mesh are shown in Figure 8.

Table 2. Configurations of the PMMA specimen.

Specimen	a	b
Case I	25.4	152.4
Case II	38.1	127

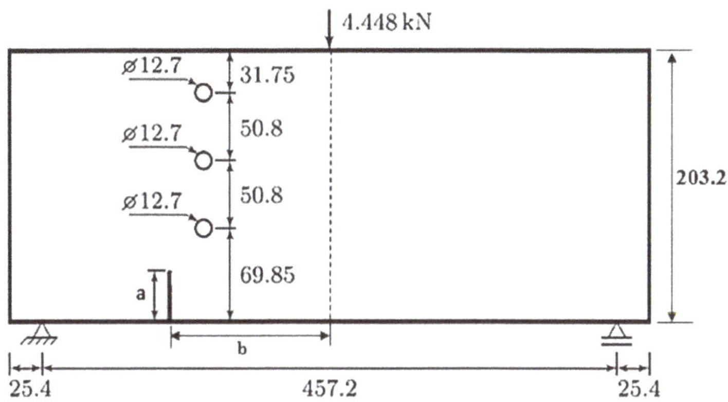

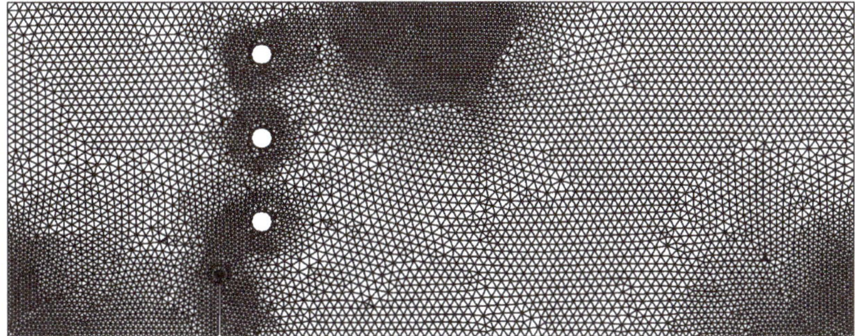

Figure 8. Problem statement for the PMMA specimen (dimensions in mm) and initial adaptive dens mesh.

3.2.1. Case I

The simulated crack growth for this specimen moved between the bottom and mid-hole and reached the mid-hole on the right side. It presented a significant increase in the KII component of the shear stress intensity factor across the cracks, which forced the step-sizes of the crack to be shortened. The findings of the crack trajectory during propagation were excellently consistent with the experimental results of the crack trajectory [32], the numerical results obtained by [33] using A polygonal extended finite element method (XFEM) with numerical integration for linear elastic fracture mechanics, the XFEM results using ABAQUS software obtained by obtained by [34], and with the numerical results using the coupled extended meshfree–smoothed meshfree method presented by [35], as shown in Figure 9a–e, respectively. The maximum principal stress distribution is shown in Figure 10.

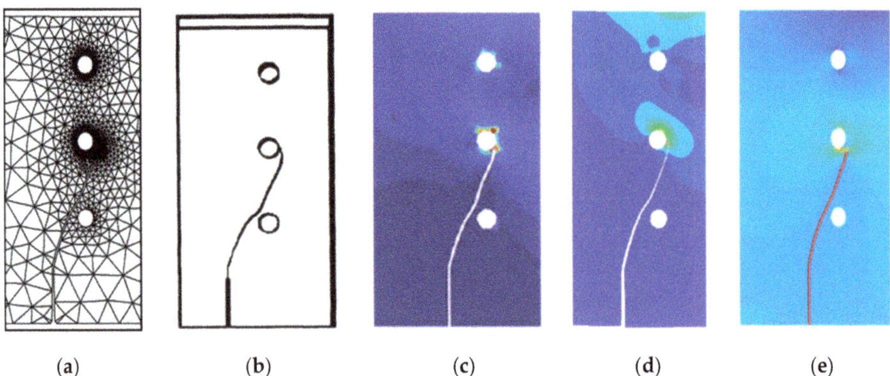

Figure 9. Final crack growth path for case I: (**a**) present study, (**b**) experimental results [32], (**c**) [33] with permission of Elsevier 2019, (**d**) [34] with permission of Elsevier 2018, and (**e**) [35] with permission of Elsevier 2020.

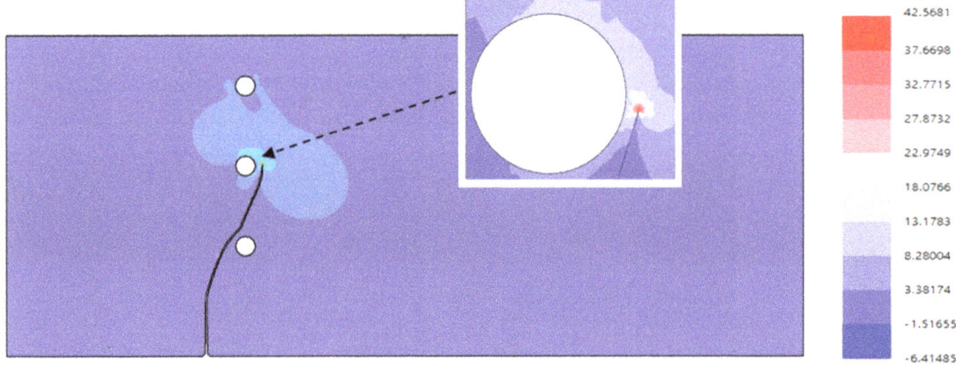

Figure 10. Maximum principal stress distribution of case I for the PMMA specimen.

The results of this simulation were compared with those from XFEM using the smooth nodal stress technique by Peng et al. 2017, as shown in Figure 11, with good agreement. It was found that as the crack approached the hole, the SIFs appeared to change to a greater amplitude. The predicted fatigue life for this specimen was compared to the analytical results calculated by [36] using Paris and Walker models, as shown in Figure 12, with good agreement.

3.2.2. Case II

According to Table 2, the differences between this case and the previous case were the initial crack length and its position from the mid-span, which were 38.1 mm and 127 mm, respectively. The crack moved above the lower hole in this specimen and stopped at the central hole from the left. The results of the crack trajectory during propagation were excellently close to the experimental results of the crack trajectory obtained by [32], XFEM results using ABAQUS software obtained by [34], as well as the numerical results obtained by [35] using the coupled extended meshfree–smoothed meshfree method, as shown in Figure 13a–d, respectively. The distribution of the von Mises stress is shown in Figure 14.

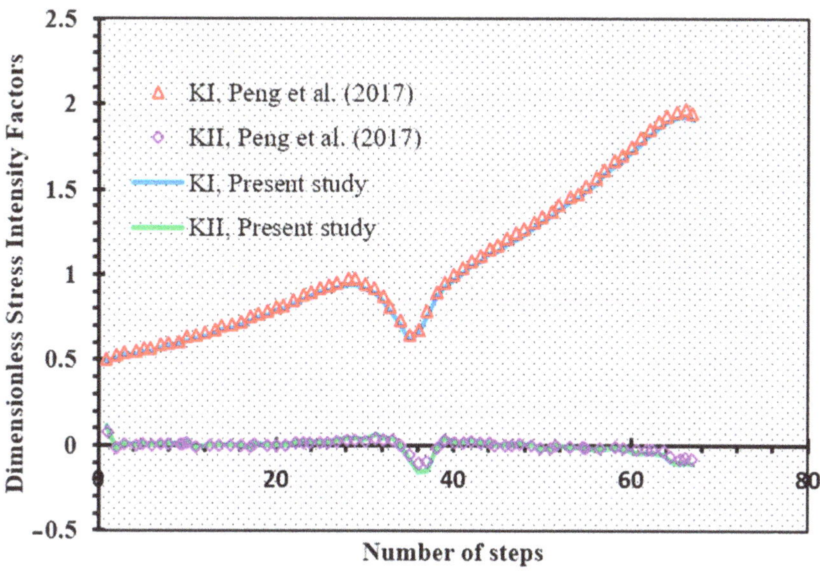

Figure 11. Non-dimensional stress intensity factors for case I of the PMMA specimen.

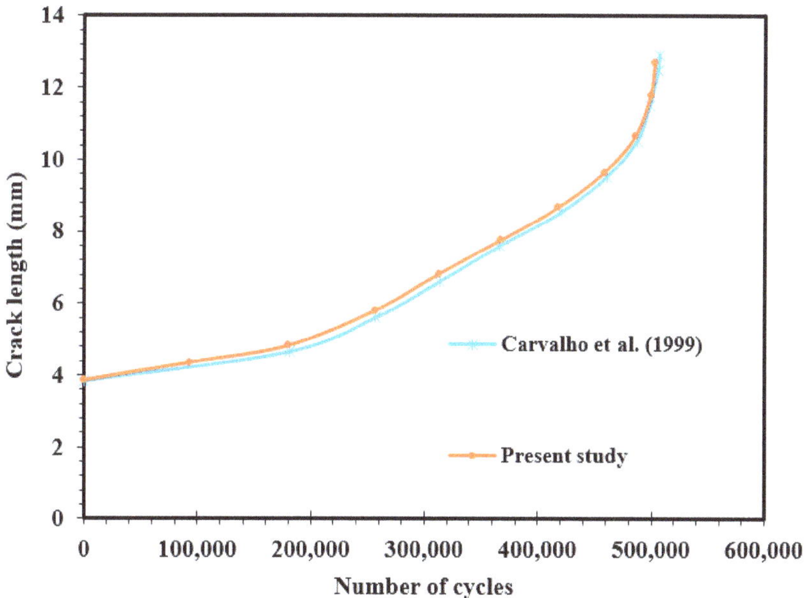

Figure 12. Predicted crack growth path for case I of the PMMA specimen.

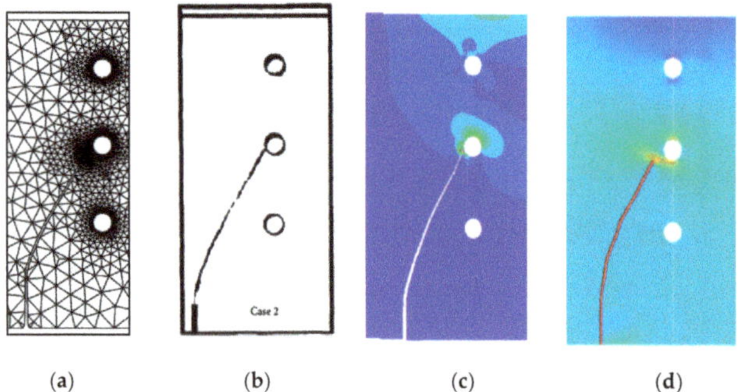

(**a**)   (**b**)   (**c**)   (**d**)

Figure 13. Final crack growth path for case I: (**a**) present study, (**b**) experimental results [32], (**c**) [34] with permission of Elsevier 2018, and (**e**) [35] with permission of Elsevier 2020.

Figure 14. Von Mises stress distribution of case II for the PMMA specimen.

The findings of the study for the dimensionless stress factor were compared with those achieved in XFEM with the smooth nodal stress system [37], as seen in Figure 15, with identical results.

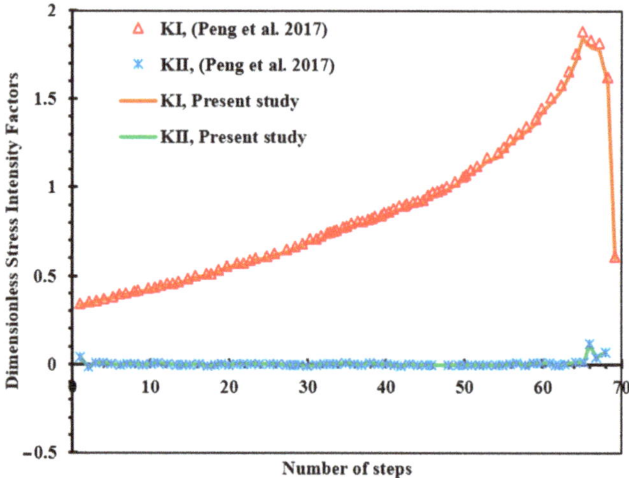

Figure 15. Comparison of the dimensionless stress intensity factors for case II for the PMMA specimen.

4. Conclusions

The results of the developed program simulation were compared with experimental and numerical data for the two internal non-colinear cracks and the three-point bending beam with three holes with two different configurations. The developed program combines the adaptive mesh refinement with increasing mesh density in the required area only in order to reduce the computational time while increasing the solution accuracy. The norm stress error is taken as a posterior estimator for the h-type adaptive refinement. With this series of simulations, the capability of the developed program was demonstrated to accurately predict the crack path trajectory, stress intensity factors, and fatigue life under constant amplitude loading. In these simulation sequences, holes act as a crack stopper and attract a crack trajectory to growth. Such findings support that the algorithm can be used to identify crack-stopping holes used in damage tolerance designs.

Author Contributions: Conceptualization, A.M.A.; methodology, A.M.A.; software, A.M.A.; validation, A.M.A., and Y.A.F.; formal analysis, A.M.A.; investigation, A.M.A. and Y.A.F.; resources, A.M.A.; data curation, A.M.A. and Y.A.F.; writing—original draft preparation, A.M.A.; writing—review and editing, A.M.A. and Y.A.F.; visualization, A.M.A. and Y.A.F.; supervision, A.M.A.; project administration, A.M.A. and Y.A.F.; funding acquisition, A.M.A. and Y.A.F. Both authors have read and agreed to the published version of the manuscript.

Funding: This research received no external funding.

Institutional Review Board Statement: Not applicable.

Informed Consent Statement: Informed consent was obtained from all subjects involved in the study.

Data Availability Statement: The data presented in this study are available in the article.

Conflicts of Interest: The authors declare no conflict of interest.

References

1. Liu, A.F. *Mechanics and Mechanisms of Fracture: An Introduction*; ASM International: Geauga County, OH, USA, 2005.
2. ANSYS. Academic Research Mechanical, Release 19.2, Help System. In *Coupled Field Analysis Guide*; ANSYS, Inc.: Canonsburg, PA, USA, 2020.
3. Manual, A.U. *Abaqus User Manual*; Abacus: Waltham, MA, USA, 2020.
4. Lebaillif, D.; Recho, N. Brittle and ductile crack propagation using automatic finite element crack box technique. *Eng. Fract. Mech.* **2007**, *74*, 1810–1824. [CrossRef]
5. Duflot, M.; Nguyen-Dang, H. Fatigue crack growth analysis by an enriched meshless method. *J. Comput. Appl. Math.* **2004**, *168*, 155–164. [CrossRef]
6. Yan, X. Automated simulation of fatigue crack propagation for two-dimensional linear elastic fracture mechanics problems by boundary element method. *Eng. Fract. Mech.* **2007**, *74*, 2225–2246. [CrossRef]
7. Zhao, T.; Zhang, J.; Jiang, Y. A study of fatigue crack growth of 7075-T651 aluminum alloy. *Int. J. Fatigue* **2008**, *30*, 1169–1180. [CrossRef]
8. Miranda, A.C.D.O.; Meggiolaro, M.A.; Martha, L.F.; Castro, J. Stress intensity factor predictions: Comparison and round-off error. *Comput. Mater. Sci.* **2012**, *53*, 354–358. [CrossRef]
9. Ingraffea, A.R.; de Borst, R. Computational Fracture mechanics. In *Encyclopedia of Computational Mechanics*, 2nd ed.; John Wiley & Sons, Ltd.: West Sussex, UK, 2017; pp. 1–26.
10. Liu, Z.; Wang, X.; Miller, R.E.; Hu, J.; Chen, X. Ductile fracture properties of 16MND5 bainitic forging steel under different in-plane and out-of-plane constraint conditions: Experiments and predictions. *Eng. Fract. Mech.* **2021**, *241*, 107359. [CrossRef]
11. Alshoaibi, A.M. Finite element procedures for the numerical simulation of fatigue crack propagation under mixed mode loading. *Struct. Eng. Mech.* **2010**, *35*, 283–299. [CrossRef]
12. Alshoaibi, A.M. An Adaptive Finite Element Framework for Fatigue Crack Propagation under Constant Amplitude Loading. *Int. J. Appl. Sci. Eng.* **2015**, *13*, 261–270.
13. Alshoaibi, A. Finite element-based model for crack propagation in linear elastic materials. *Eng. Solid Mech.* **2020**, *8*, 131–142. [CrossRef]
14. Alshoaibi, A.M.; Hadi, M.; Ariffin, A. Two-dimensional numerical estimation of stress intensity factors and crack propagation in linear elastic Analysis. *J. Struct. Durab. Health Monit.* **2007**, *3*, 15–27.
15. Fageehi, Y.A.; Alshoaibi, A.M. Nonplanar Crack Growth Simulation of Multiple Cracks Using Finite Element Method. *Adv. Mater. Sci. Eng.* **2020**, *2020*, 1–12. [CrossRef]

16. Alshoaibi, A.M.; Ariffin, A. Finite element modeling of fatigue crack propagation using a self adaptive mesh strategy. *Int. Rev. Aerosp. Eng.* **2015**, *8*, 209–215. [CrossRef]
17. Alshoaibi, A.M.; Ariffin, A. Fatigue life and crack path prediction in 2D structural components using an adaptive finite element strategy. *Int. J. Mech. Mater. Eng.* **2008**, *3*, 97–104.
18. Alshoaibi, A.M.; Ariffin, A.K. Finite element simulation of stress intensity factors in elastic-plastic crack growth. *J. Zhejiang Univ. Sci. A* **2006**, *7*, 1336–1342. [CrossRef]
19. Alshoaibi, A.M. A two dimensional Simulation of crack propagation using Adaptive Finite Element Analysis. *J. Comput. Appl. Mech.* **2018**, *49*, 335–341.
20. Alshoaibi, A.M.; Fageehi, Y.A. Fageehi, 2D finite element simulation of mixed mode fatigue crack propagation for CTS specimen. *J. Mater. Res. Technol.* **2020**, *9*, 7850–7861. [CrossRef]
21. Bashiri, A.H.; Alshoaibi, A.M. Adaptive Finite Element Prediction of Fatigue Life and Crack Path in 2D Structural Components. *Metals* **2020**, *10*, 1316. [CrossRef]
22. Barsoum, R.S. Application of quadratic isoparametric finite elements in linear fracture mechanics. *Int. J. Fract.* **1974**, *10*, 603–605. [CrossRef]
23. Henshell, R.D.; Shaw, K.G. Crack tip finite elements are unnecessary. *Int. J. Numer. Methods Eng.* **1975**, *9*, 495–507. [CrossRef]
24. Phongthanapanich, S.; Dechaumphai, P. Adaptive Delaunay triangulation with object-oriented programming for crack propagation analysis. *Finite Elements Anal. Des.* **2004**, *40*, 1753–1771. [CrossRef]
25. Erdogan, F.; Sih, G.C. On the crack extension in plates under plane loading and transverse shear. *J. Basic Eng.* **1963**, *85*, 519–525. [CrossRef]
26. Zienkiewicz, O.; Taylor, R.; Zhu, J. *The Finite Element Method: Its Basis and Fundamentals*; Elsevier: Amsterdam, The Netherlands, 2005.
27. Cavendish, J.C. Automatic triangulation of arbitrary planar domains for the finite element method. *Int. J. Numer. Methods Eng.* **1974**, *8*, 679–696. [CrossRef]
28. Andersen, M.R. *Fatigue Crack Initiation and Growth in Ship Structures*; Department of Naval Architecture and Offshore Engineering, Technical University of Denmark: Lyngby, Denmark, 1998.
29. Surendran, M.; Bordas, S.; Palani, G.; Bordas, S. Linear smoothed extended finite element method for fatigue crack growth simulations. *Eng. Fract. Mech.* **2019**, *206*, 551–564. [CrossRef]
30. Tu, S.; Cai, R. A coupling of boundary elements and singular integral equation for the solution of the fatigue cracked body. *Stress Anal.* **1993**, *3*, 239–247.
31. Bittencourt, T.; Wawrzynek, P.; Ingraffea, A.; Sousa, J. Quasi-automatic simulation of crack propagation for 2D LEFM problems. *Eng. Fract. Mech.* **1996**, *55*, 321–334. [CrossRef]
32. Ingraffea, A.R.; Grigoriu, M. *Probabilistic Fracture Mechanics: A Validation of Predictive Capability*; Cornell University Ithaca Ny Depterment of Structural Engineering: Ithaca, NY, USA, 1990.
33. Huynh, H.D.; Nguyen, M.N.; Cusatis, G.; Tanaka, S.; Bui, T.Q. A polygonal XFEM with new numerical integration for linear elastic fracture mechanics. *Eng. Fract. Mech.* **2019**, *213*, 241–263. [CrossRef]
34. Dirik, H.; Yalçinkaya, T. Crack path and life prediction under mixed mode cyclic variable amplitude loading through XFEM. *Int. J. Fatigue* **2018**, *114*, 34–50. [CrossRef]
35. Ma, W.; Liu, G.; Wang, W. A coupled extended meshfree–Smoothed meshfree method for crack growth simulation. *Theor. Appl. Fract. Mech.* **2020**, *107*, 102572. [CrossRef]
36. Carvalho, C.V.; de Araújo, T.D.P.; Cavalcante, J.B.; Martha, L.F.; Bittencourt, T. Automatic Fatigue Crack Propagation Using a Self-Adaptative Strategy. In Proceedings of the PACAM VI—Sixth Pan-American Congress of Applied Mechanics, Rio de Janeiro, Brazial; 1999. Available online: file:///C:/Users/MDPI/AppData/Local/Temp/Automatic_fatigue_crack_propagation_using_a_self-a.pdf (accessed on 15 October 2020).
37. Peng, X.; Kulasegaram, S.; Wu, S.; Bordas, S. An extended finite element method (XFEM) for linear elastic fracture with smooth nodal stress. *Comput. Struct.* **2017**, *179*, 48–63. [CrossRef]

Article

Analysis and Optimization of Tooth Surface Contact Stress of Gears with Tooth Profile Deviations, Meshing Errors and Lead Crowning Modifications Based on Finite Element Method and Taguchi Method

Qiang Li [1,2,*] and Liyang Xie [1,2]

1   School of Mechanical Engineering and Automation, Northeastern University, Shenyang 110819, China; lyxieneu@163.com
2   Key Laboratory of Vibration and Control of Aero-Propulsion System Ministry of Education, Northeastern University, Shenyang 110819, China
*   Correspondence: 1410093@stu.neu.edu.cn; Tel.: +86-138-981-35697

Received: 6 September 2020; Accepted: 12 October 2020; Published: 14 October 2020

Abstract: Based on the three-dimensional (3D) finite element method (FEM) and Taguchi method (TM), this paper analyzes the tooth surface contact stress (TSCS) of spur gears with three different influence factors: tooth profile deviations (TPD), meshing errors (ME) and lead crowning modifications (LCM), especially researching and analyzing the interactions between TPD, ME and LCM and their degree of influence on the TSCS. In this paper, firstly, a 3D FEM model of one pair of engaged teeth is modeled and the mesh of the contact area is refined by FEM software. In the model, the refined area mesh and the non-refined area mesh are connected by multi-point constraint (MPC); at the same time, in order to save the time of the FEM solution on the premise of ensuring the solution's accuracy, the reasonable size of the refined area is studied and confirmed. Secondly, the TSCS analyses of gears with one single influence factor (other factors are all ideal) are carried out. By inputting the values of different levels of one single factor into the FEM model, especially using the real measurement data of TPD, and conducting the TSCS analysis under different torques, the influence degree of one single factor on TSCS is discussed by comparing the ideal model, and it is found that when the influence factors exist alone, each factor has a great influence on the TSCS. Finally, through TM, an orthogonal test is designed for the three influence factors. According to the test results, the interactions between the influence factors and the influence degree of the factors on the TSCS are analyzed when the three factors exist on the gear at the same time, and it is found that the TPD has the greatest influence on the TSCS, followed by the lead crowning modified quantity. The ME is relatively much small, and there is obvious interaction between ME and LCM. In addition, the optimal combination of factor levels is determined, and compared with the original combination of a gear factory, we see that the contact fatigue performance of the gear with the optimal combination is much better. The research of this paper has a certain reference significance for the control of TPD, ME and LCM when machining and assembling the gears.

Keywords: finite element method; Taguchi method; tooth surface contact stress; tooth profile deviations; meshing errors; lead crowning modifications

---

1. Introduction

Gear transmission is the most important and widely used transmission in mechanical transmission. In particular, long-life precision gears are widely used in mechanical transmission systems with high reliability requirements, such as wind turbine gearboxes [1] and aero engines [2]. As the requirements for long-life precision gears are increasing, the requirements of machining accuracy, assembly accuracy and gear tooth modification are also growing, and these factors have a great influence on the tooth surface contact stress (TSCS) of gears.

At present, the main calculation methods of the TSCS include certain international standards such as ISO [3,4]. Most of these standard calculations are based on experience and some experimental results. However, when there are different tooth profile deviations (TPD), meshing errors (ME) and lead crowning modifications (LCM) on the tooth surface, since the influence coefficient in the calculation formula is derived from research results and field experience, and the determination of some coefficients is affected by many factors, it is difficult to determine the value of the influence coefficients. In addition, the interaction between each influence coefficient is not reflected in the formula. Therefore, the above methods cannot perform accurate calculations [5]. Therefore, it is very important to accurately analyze and calculate the TSCS of gears with the above influence factors, especially when the three factors exist at the same time.

At present, there are a lot of studies on the influence of TPD, ME and LCM on gear performance. Sennba [6] studied the influence of LCM gears and gear shaft misalignment on TSCS and tooth root bending stress through a large number of tests as early as 1974. Ottewill et al. [7] derived an equation of motion incorporating an error function and losses at the mesh interface to study the effect of TPD on the rattling of idling gears. They compared the theoretical and experimental trajectories by way of time domain plots as well as via contour plots, and for most profile error functions, good agreement was achieved between the model and experimental data. The results showed that the TPD weakened the contact strength and bending strength of gears, and intensified the vibration and noise of gears [7]. Ma et al. [8] introduced a method to analyze the dynamic load of the planetary gears in the gearbox of a PT6 turboprop engine, and a rigorous dynamic analysis, which included the effects of nonlinear tooth stiffnesses, ring gear flexibility, gear errors and misalignments, was necessary to determine dynamic tooth loads and the load sharing among the planets. The results were presented from sample calculations for a typical gear stage [8]. Bodas et al. [9] used an advanced contact mechanics model of a planetary gear set to study the effects of a series of manufacturing and assembly errors on the load distribution between planetary gears. Three different groups of errors were considered: (i) time-invariant, assembly-independent errors, (ii) time-invariant, assembly-dependent errors, and (iii) time-varying, assembly-dependent errors. With such errors present, the planet load sharing characteristics of an n-planet system (n = 3 to 6) were investigated for different piloting configurations under both static and dynamic conditions [9]. In view of the shortcomings of traditional analysis methods (the complicated integral equations make it difficult to instantly obtain proper results for some tooth flanks distorted by heat treatment, and the repetition calculation may not converge, especially in light load conditions), Miyosh et al. [10] proposed a new composite analysis method which can quickly calculate the contact load distribution on the tooth surface of helical gears under any load condition. Although a large number of scholars have carried out experimental research and theoretical research, these studies have not put forward a method or model that can efficiently and accurately calculate the tooth stress when the above three factors exist at the same time, especially for the influence of these factors on the TSCS when there are different levels of these factors.

In recent years, more and more scholars have used the finite element method (FEM) to model gears, and then conduct the simulation and analysis of the gear performance, and confirm that the FEM is a very effective method for the simulation analysis of gears. Li [5] has used the FEM to calculate the TSCS and the tooth root stress with manufacturing errors, tooth profile modifications and assembly errors. By comparing the test results of other scholars and their own test results, it was found that the calculation results of the FEM including ME and LCM are relatively consistent with the test results, which indicates the accuracy of the FEM. It also was found that TPD, ME and LCM exert great effects on the TSCS of the gears. The TSCS and the tooth root stress of the same pair of gears were also calculated by ISO (International Organization for Standardization) and JGMA (Japan Gear Manufacture Association) standards for comparing with the FEM results, and the accuracy and efficiency of the finite element model have been verified [5]. However, the scholar only studied the influence of a single influence factor on TSCS under a single torque load through the finite element method, and did not study the influence of different levels of influence factors on the TSCS under different torques. In addition, although the scholar conducted a simple finite element analysis when all factors existed at the same time, the influence degree of each factor on the TSCS and the interaction between the factors have not been studied and analyzed. Wang et al. [11] used the FEM to analyze the TSCS and transmission error of the gears with tooth profile errors, and the results showed that mesh stiffness decreases and loaded transmission error and the maximum tooth contact stress grow when increasing the tooth lead crown relief. Lin et al. [12] used the FEM to model the machining errors, assembly errors and tooth modifications of the gear transmission system to obtain the static transmission error, so as to analyze the coupled transmission error of a helical gear system. Although there are a lot of FEM studies on the performance of gears with different influencing factors, the main highlight is the influence of one single factor on TSCS, and this does not consider the interaction between various factors, especially the influence of these factors on the gear performance when they exist at the same time.

In addition, most studies often only study the influence of one single precision grade of TPD on gear performance, or assume that the distribution form of TPD on the tooth surface is a sine function, which does not represent the influence of the actual TPD of different precision grades on the gear performance.

When researching the interaction of various factors and the influence degree of various factors on TSCS when they exist at the same time, this paper uses the Taguchi method (TM). The TM has been widely used for the design and analysis of various engineering fields, such as gear manufacturing processes [13], chemical industry [14], soil [15] and energy [16], and it has been demonstrated to be a powerful tool due to its simplicity and robustness [17]. Through the TM, within the specified level range, the optimal level combination of influence factors can also be obtained to determine the optimal TPD grade, ME and LCM quantity in order to obtain a relatively minimum TSCS.

In summary, compared with the experimental method and theoretical method, the FEM can obtain the TSCS more effectively, accurately and intuitively, such that a 3D FEM model of one pair of engaged teeth is modeled and the mesh of the contact area is refined by FEM software in this paper. Then, by inputting the values of different levels of one single factor into the FEM model, especially using the real measurement data of TPD, and conducting the TSCS analysis under different torques, the influence degree of one single factor on TSCS is discussed by comparing the ideal model. In view of the fact that other scholars rarely study the degree of influence of each factor on TSCS and the interaction between them when all factors exist at the same time, in this paper, through the TM, the interactions between the influence factors and the influence degree of the factors on the TSCS are studied when the three factors exist on the gear at the same time, and the optimal combination of factor levels can also be obtained.

The general analysis steps of this paper are shown in Figure 1.

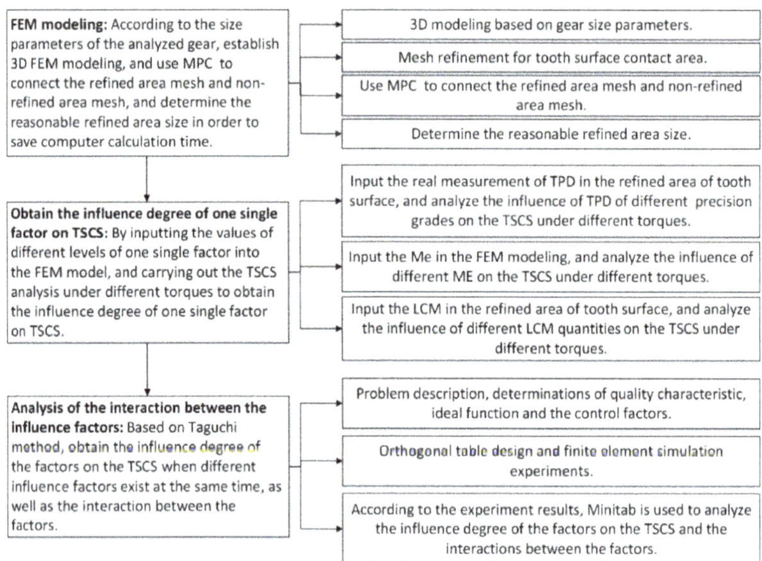

Figure 1. The general analysis steps.

2. FEM Modeling

2.1. Determination of Gear Engagement Position

In this paper, a pair of gears with the same parameters are selected for engagement. The basic size parameters of the gears are shown in Table 1. Through calculation, the coincidence degree of the engaged gear is 1.56 so there are engagement positions of a single pair of teeth in the entire engagement circle.

Table 1. The basic size parameters of the gears.

Normal Modulus	Tooth Number	Pressure Angle	Addendum Coefficient	Clearance Coefficient	Helix Angle	Tooth Width	Modification Coefficient	Poisson Ratio	Elastic Modulus
2 mm	20	20°	1	0.25	0°	5 mm	0	0.25	2.07×10^5 MPa

As it can be seen from the literature [18] and Figure 2, for ideal gears, when the engagement positions are single-pair tooth contact positions (positions 16–27), the single pair of teeth bears all the applied load, that is to say, for one period of tooth engagement, the maximum TSCS occurs in the positions 16–27. As such, this paper defines the engagement position of the TSCS analysis at the position of the gear pitch circle, and the TPD, ME and LCM of the tooth surface are input and adjusted based on this engagement position, so as to research and analyze the influence of different factors on the TSCS.

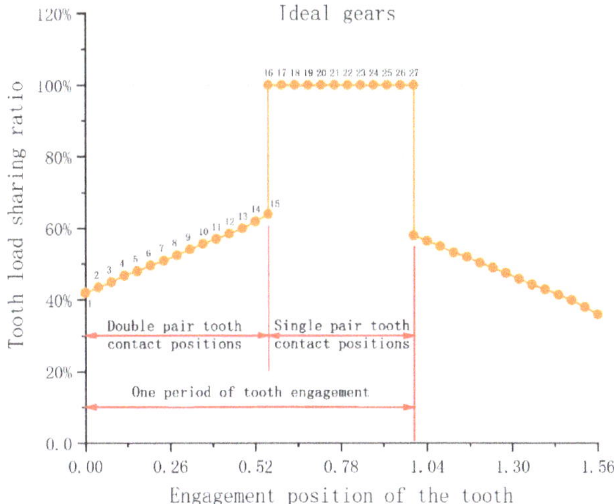

Figure 2. The load-sharing ratios of the ideal gears.

2.2. Multi-Point Constraint (MPC)

When the gears are engaged, the stress field gradient near the contact line is very large. The contact area needs high-density mesh to capture the contact state. The smaller the element size of the FEM model is, the smaller the stress difference between the elements is, and the higher the solution accuracy of the model is, but the longer the solution time of the model is. In order to balance the contradiction between solution accuracy and solution time, it is necessary to determine a reasonable mesh density transition boundary.

In order to ensure the solution accuracy of the FEM, it is necessary to refine the mesh of the contact area. In this paper, MPC is used to connect the refined mesh with the non-refined mesh. As shown in Figure 3, the orange part is the refined area of the tooth mesh. Partial FEM parameters are shown in Table 2.

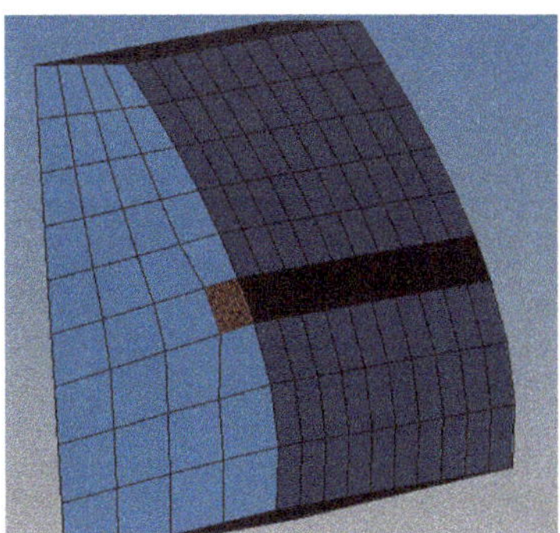

Figure 3. The refined area of the tooth mesh.

Table 2. The finite element method (FEM) parameters.

Type of Main Elements	Contact Surface Element	Target Surface Element	Friction Coefficient	Material Density
Solid185	Conta173	Targe170	0.1	7.8×10^{-9} t/mm^3

MPC, that is, multi-point constraint, establishes a multi-point constraint relationship. Through MPC, different meshes can be connected. If the geometry is not connected in the topology, different geometric parts can be meshed respectively, and then the FEM models can be connected with the MPC. A simple example is shown in Figure 4; Figure 4a shows the MPC connection of refined mesh and non-refined mesh, while Figure 4b shows no MPC connection and the entire mesh has been refined. Both models are solved using SHELL181 element, in which the shell element thickness is 0.01 m, the elastic modulus $E = 2.1 \times 10^5$ MPa, and the Poisson's ratio $\mu = 0.3$. Figure 4c shows the loading type and boundary conditions, and the two models have the same type of boundary conditions and linear pressure loads, in which the linear pressure is 10 N/m. Figure 4d,e show the analysis results of FEM, and the maximum stress on both models is 6.63 MPa. According to the stress nephogram and solution results, for the FEM model of MPC connection, when the size of the refined area is large enough, that is to say, when the distance between the MPC connection position and the stress analysis position is far enough, the solution results are basically consistent with the results of the entire refined mesh model. Therefore, in this paper, the mesh of the refined area and non-refined area is connected by the MPC, and in order to save the time of FEM analysis, and at the same time ensure the accuracy of the solution, the optimal size of the refined area will be studied and analyzed to determine the mesh transition boundary position.

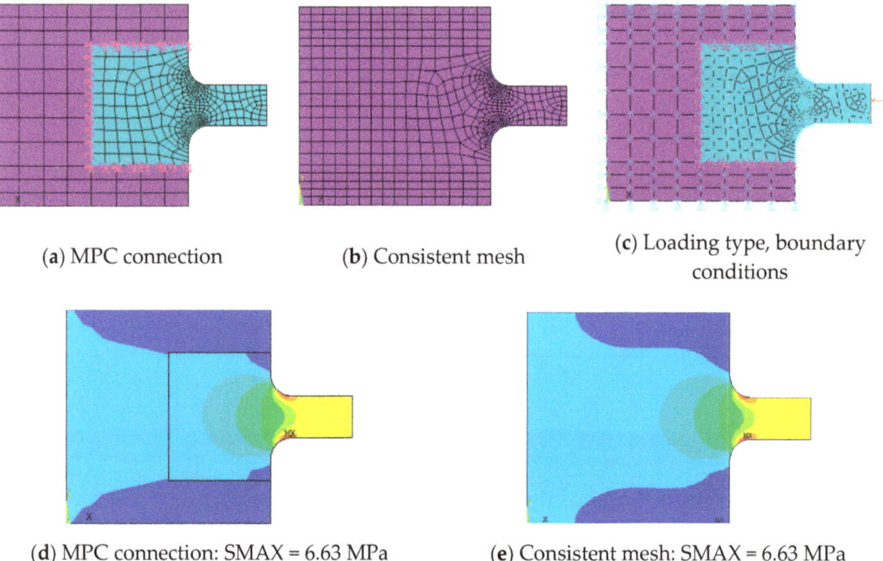

<table>
<tr><td>(a) MPC connection</td><td>(b) Consistent mesh</td><td>(c) Loading type, boundary conditions</td></tr>
<tr><td colspan="2" align="center">(d) MPC connection: SMAX = 6.63 MPa</td><td>(e) Consistent mesh: SMAX = 6.63 MPa</td></tr>
</table>

Figure 4. Multi-point constraint (MPC) connection.

2.3. Determination of Mesh Transition Boundary Position

2.3.1. Hertz Contact Theory

In order to determine the optimal position of the mesh transition boundary, it is necessary to establish different-sized refined areas to verify the accuracy of the model. The engagement of involute cylindrical gears is similar to the contact between two cylinders, so the accuracy of the FEM solution can be verified by Hertz contact theory [19]. Figure 5 is a Hertz contact model, where F is the normal force applied, σ_{Hmax} is the maximum contact stress on the contact surface, b is the half width of the contact zone, a is the contact length, and ρ_1 and ρ_2 are the curvature radius of the two cylinders. σ_{Hmax} and b can be calculated by Equations (1) and (2).

$$\sigma_{Hmax} = \sqrt{\frac{F}{\pi \times a} \times \frac{\left(\frac{1}{\rho_1} + \frac{1}{\rho_2}\right)}{\left(\frac{1-\mu_1^2}{E_1} + \frac{1-\mu_2^2}{E_2}\right)}} \tag{1}$$

$$b = \sqrt{\frac{4 \times F}{\pi \times a} \times \frac{\left(\frac{1-\mu_1^2}{E_1} + \frac{1-\mu_2^2}{E_2}\right)}{\left(\frac{1}{\rho_1} + \frac{1}{\rho_2}\right)}} \tag{2}$$

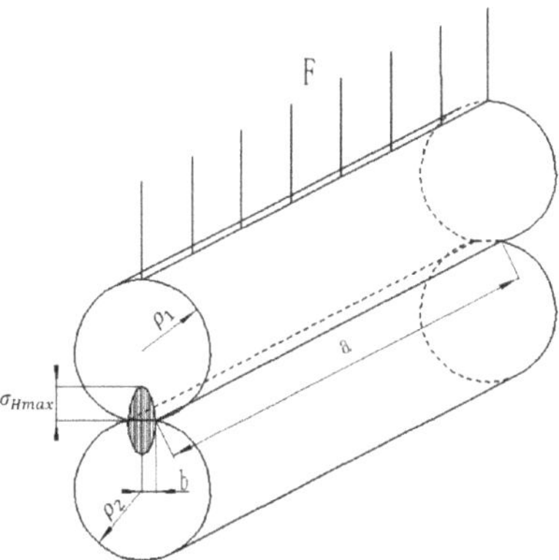

Figure 5. Hertz contact model.

In Equations (1) and (2), μ_1 and μ_2 are the Poisson's ratio of the materials of cylinder 1 and cylinder 2, and E_1 and E_2 are the elastic modulus of the materials of cylinder 1 and cylinder 2. Treat the cylindrical contact as the engaged contact of two gear teeth; a is equivalent to the tooth width, and F is equivalent to the normal force of the tooth surface contact. In Equations (3)–(5), F_t is the circumferential force produced by torque, α is the pressure angle, and d_1 and d_2 are the diameters of the pitch circles.

$$F = \frac{F_t}{cos\alpha} = \frac{2 \times T}{d_1 \times cos\alpha} \tag{3}$$

$$\rho_1 = \frac{d_1 \times sin\alpha}{2} \tag{4}$$

$$\rho_2 = \frac{d_2 \times sin\alpha}{2} \tag{5}$$

The maximum torque T applied to the engaged gear is defined as 25 N·m. Combining Equations (1)–(5), the TSCS of the engaged gear is calculated to be 1653.75 MPa, and the half width of the contact zone is 0.10 mm. The TSCS has basically reached the contact fatigue limit of carburized alloy steel gears with high material quality and heat treatment quality. Therefore, in this paper, the maximum torque applied to the gears will not exceed 25 N·m. If the size of the refined area determined under this torque can meet the solution accuracy, the solution accuracy can also be guaranteed when the torque is less than 25 N·m.

2.3.2. Determination of Mesh Transition Boundary Position

As shown in Figure 6, take the length of b as the basic unit of the refined area size, and set the MPC boundary of the contact area as $w \times h$. Firstly, set $w \times h$ as $8b \times 8b$ to refine the mesh, and the FEM model of a pair of ideal teeth is shown in Figure 7.

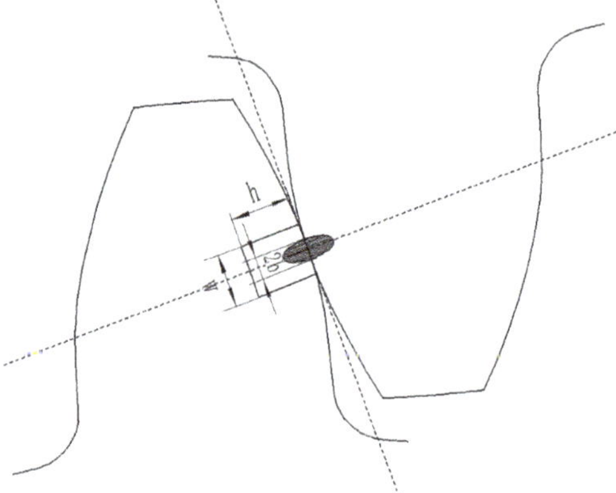

Figure 6. MPC boundary.

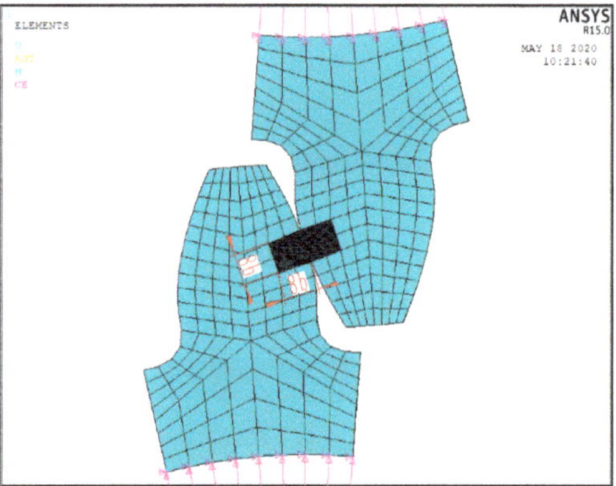

Figure 7. Finite element method (FEM) model.

Apply a torque of 25 Nm to the FEM model and conduct FEM solution. The Von Mises stress nephogram is shown in Figure 8. It can be seen from Figure 8 that the stress gradient of the gear at the engagement position is very large, but the stress decreases sharply at a certain distance from the maximum stress position, and at a position less than $4b$ below the tooth surface, the stress gradient is very gentle. Therefore, when the size of the refined area is $w \times h = 8b \times 8b$, the influence of the MPC on the solution accuracy of the model can be ignored. It can be seen from Figure 9 that the maximum TSCS solved by the FEM model is 1676.23 MPa, and the relative error is only 1.36% when compared with the 1653.75 MPa calculated by the Hertz contact theory.

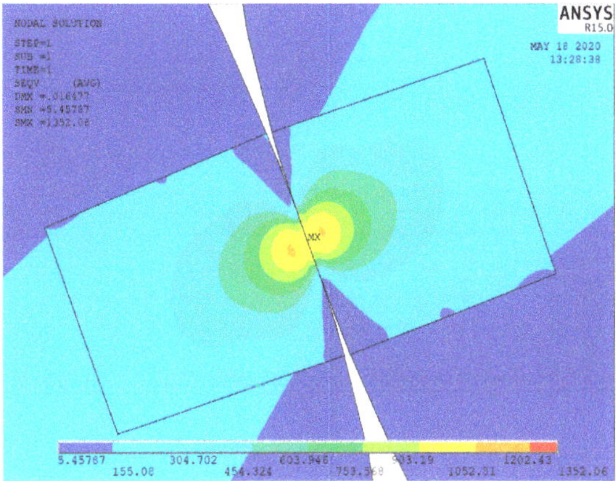

Figure 8. The Von Mises stress nephogram of solution result.

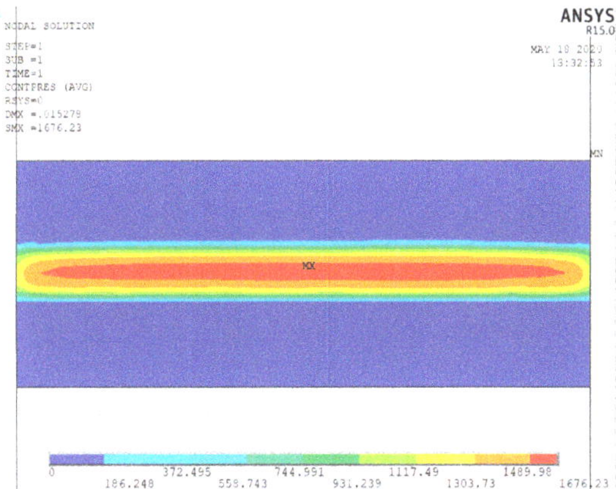

Figure 9. The maximum tooth surface contact stress (TSCS) solved by the FEM model.

When the size of the refined area is $w \times h = 8b \times 8b$, the solution accuracy meets the requirements, so the size of the refined area is further reduced. Figure 10 shows the Von Mises stress nephograms corresponding to the refined areas of different sizes, and Table 3 is the corresponding maximum TSCS. It can be seen from Figure 10 and Table 3 that when $w \times h = 4b \times 4b$ and $w \times h = 4b \times 3b$, the stress gradients at the MPC connection position are relatively gentle, and the errors between the maximum TSCS and that calculated by the Hertz contact theory are within 2.5%, the solution accuracy of the model still meets the requirements. With a further reduction in the size of the refined area, when $w \times h = 4b \times 2b$ and $w \times h = 3b \times 4b$, the stress gradients at the MPC connection position are relatively large, and the errors between the maximum TSCS and that calculated by the Hertz contact theory are also relatively large.

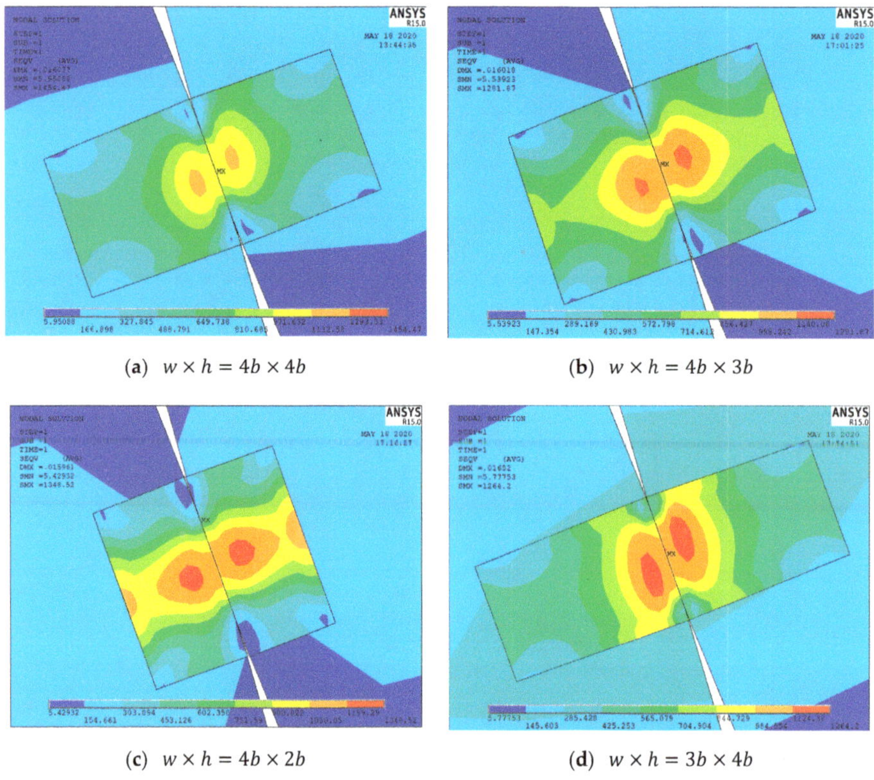

(**a**) $w \times h = 4b \times 4b$

(**b**) $w \times h = 4b \times 3b$

(**c**) $w \times h = 4b \times 2b$

(**d**) $w \times h = 3b \times 4b$

Figure 10. The Von Mises stress nephogram of solution result.

Table 3. The maximum tooth surface contact stress (TSCS) in refined areas of different sizes.

$w \times h$	The Maximum TSCS (MPa)	The Errors of TSCS	Time (min)
Hertz contact theory	1653.75		
$8b \times 8b$	1676.23	1.36%	75
$4b \times 4b$	1682.93	1.76%	9
$4b \times 3b$	1693.59	2.41%	6
$4b \times 2b$	1754.98	6.12%	4
$3b \times 4b$	1712.73	3.57%	6

In addition, the solution time shows that the larger the size of the refined area, the longer the solution time, especially when the solution time of $w \times h = 8b \times 8b$ is more than 10 times that of $w \times h = 4b \times 3b$, so if the full tooth mesh is refined, the calculation time will be much longer.

In conclusion, for the TSCS analysis of the ideal gear researched in this paper, the reasonable size of the refined area should be at least $w \times h = 4b \times 3b$, but the TSCS values of the gears with TPD, ME or LCM are much larger than that of the ideal gear. Therefore, according to the Hertz contact theory, assuming that the maximum TSCS of the gear with influence factors is 1.5 times that of the ideal gear, the contact half-width should also be increased by 1.5 times, as should the half-width of the contact zone b. Therefore, when analyzing the TSCS of the gears with influence factors, the reasonable size of the refined area should be at least $w \times h = 6b \times 4.5b = 0.60 \ \text{mm} \times 0.45 \ \text{mm}$.

In addition, referring to the calculation method of TSCS in ISO6336 [3,4], the accuracy of FEM can be further verified. According to Equation (6) and Table 4, the TSCS of the gear is 1657.28 MPa, and the TSCS error between the $w \times h = 4b \times 3b$ and Equation (6) is only 2.19%.

$$\sigma_H = Z_D Z_H Z_E Z_\varepsilon Z_\beta \sqrt{\frac{F_t}{d_1 b} \frac{u+1}{u}} \sqrt{K_A K_\gamma K_V K_{H\beta} K_{H\alpha}} \tag{6}$$

Table 4. Meanings of the symbols used in Equation (6).

Symbol	Description	Value
σ_H	Contact stress (MPa)	1657.28
Z_D	Single pair tooth contact factors for the wheel	1.0
Z_H	Zone factor	2.5
Z_E	Elasticity factor	187.5
Z_ε	Contact ratio factor	1.0
Z_β	Helix angle factor	1.0
F_t	Tangential force at the working pitch circle (N)	1250
d_1	Diameter (mm)	40
b	Face width (mm)	5
u	Gear ratio	1.0
K_A	Application factor	1.0
K_γ	Mesh load factor	1.0
K_V	Dynamic factor	1.0
$K_{H\beta}$	Face load factor	1.0
$K_{H\alpha}$	Transverse load factor	1.0

3. TSCS Analysis of Gear with TPD, ME or LCM

3.1. TSCS Analysis of Gear with TPD

All actual manufactured gears have TPD. According to the definition of TPD in the ISO 1328-1:2013 standard [20], TPD refers to the amount of actual tooth profile deviating from the designed tooth profile, which is calculated in the end plane and perpendicular to the involute tooth profile. According to the gear precision grade, gear module and pitch circle diameter, the range of the total TPD value under the precision grade is specified in the ISO standard [20].

Figure 11 shows the node distribution form of the tooth surface refined area. Each node on the tooth surface of the area is numbered, then the corresponding TPD is input to each node, and these TPD values come from the measured data of the gear factory. Figure 12 shows the tooth surface state after inputting the measured TPD of a grade 4 precision gear, and the unevenness of the tooth surface can be vaguely seen from the figure. We enlarged the tooth surface (shadow part) of the refined area in Figure 12, and show the enlarged effect in Figure 13. In Figure 13, the X axis direction is the direction from tooth root to tooth top, the Y axis direction is the direction of the tooth width, and the Z axis represents the TPD value. After enlarging the refined area, it can be clearly seen that the tooth surface is uneven.

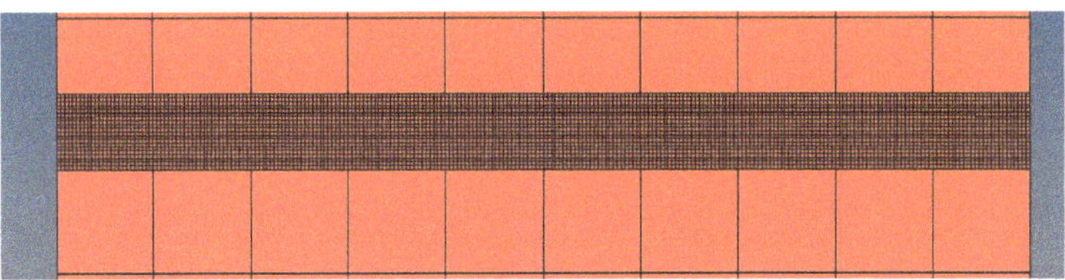

Figure 11. The node distribution form of the tooth surface's refined area.

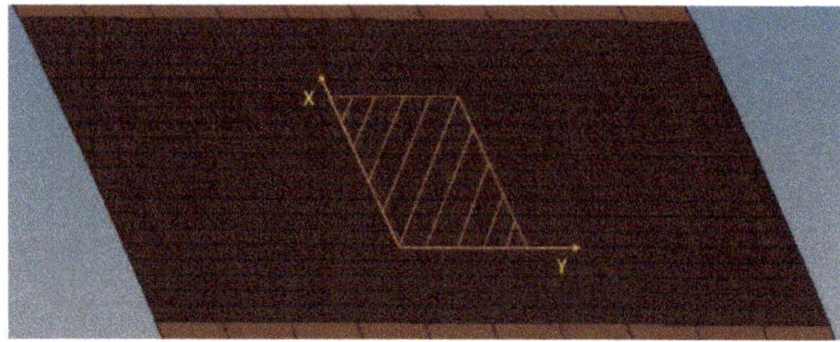

Figure 12. The tooth surface state after inputting the measured tooth profile deviations (TPD).

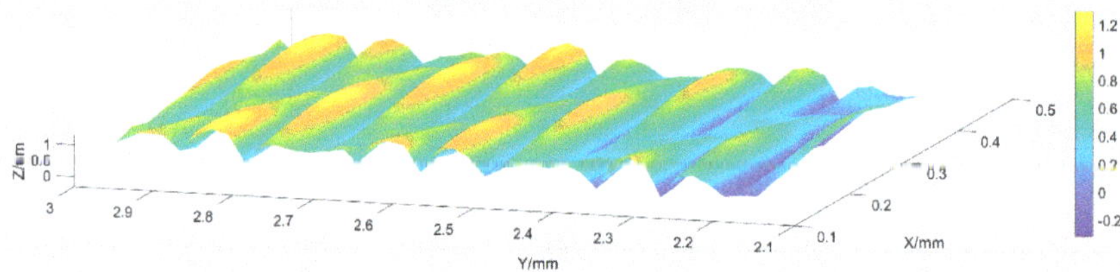

Figure 13. The enlarged effect of the tooth surface refined area with TPD.

This paper also proposes a method to obtain the TPD of other precision grade gears when the TPD value of the grade 4 precision gear is known. As shown in Table 5, according to the range of the total TPD value of each precision grade gear, it is considered that:

$$f_{ni} : f_{4i} = F_n : F_4 \tag{7}$$

where F_n is the upper limit value of the total TPD value of the grade n precision gear, and f_{ni} is the actual TPD value of each node of the grade n precision gear (i is the node number), so the TPDs of the refined areas of other precision grade gears can be obtained by Equation (7).

Table 5. The total tooth profile deviations (TPD) value of each grade precision gear.

Pitch Circle Diameter (mm)	Module (mm)	Precision Grade							
		0	1	2	3	4	5	6	7
20 < d ≤ 50	0.5 ≤ m ≤ 2	0.9 μm	1.2 μm	1.8 μm	2.5 μm	3.5 μm	5.0 μm	7.0 μm	10.0 μm

In this paper, the FEM contact analyses for grade 2, 4 and 6 precision gears with only TPD are carried out, respectively. Figure 14 is the result of the FEM solution when the torque is 10 Nm, and, from Figure 14a to Figure 14d, shows the TSCS nephograms of ideal gear, grade 2 precision gear, grade 4 precision gear and grade 6 precision gear. It can be seen that the TSCS with TPD presents irregular distribution on the tooth surface, and with the reduction of gear precision, the gradient of the local TSCS increases gradually, and the maximum TSCS also increases accordingly.

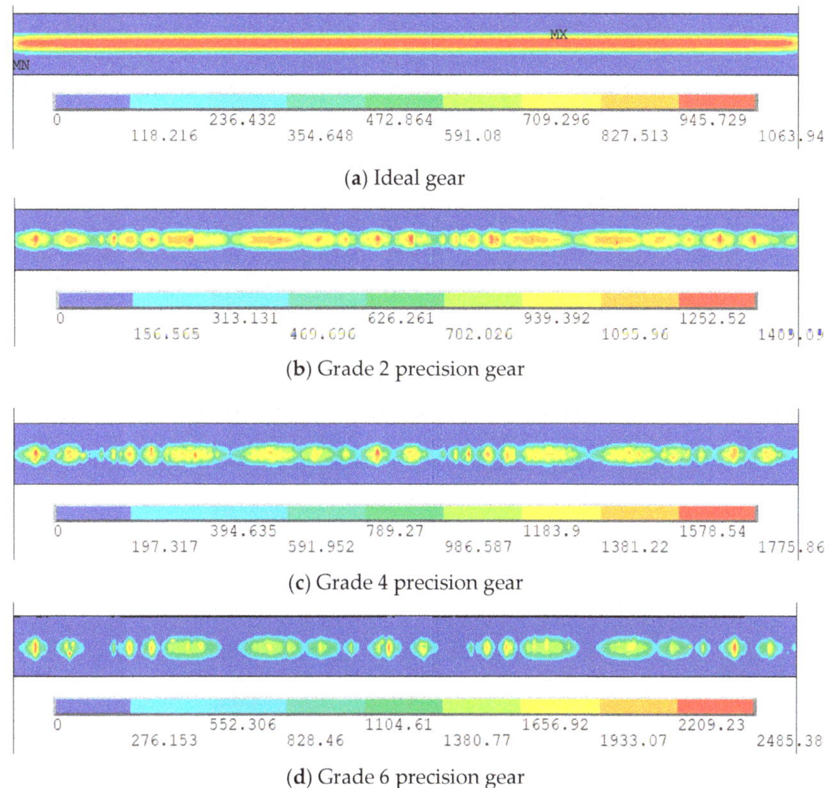

(**a**) Ideal gear

(**b**) Grade 2 precision gear

(**c**) Grade 4 precision gear

(**d**) Grade 6 precision gear

Figure 14. The TSCS nephograms of different precision grades of gears.

In this paper, the FEM contact analysis is conducted under different torques for different precision grades of gears, and the results of the maximum TSCS are compared as shown in Table 6 and Figure 15. It can be seen that the influence of TPD on the TSCS is very large, and the difference between the maximum TSCS corresponding to the different precision grades is large. At the same time, it is found that with the increase in applied load, the ratio of the maximum TSCS of each precision grade gear to the ideal gear gradually decreases.

Table 6. The results of the maximum TSCS under different torques. Unit: MPa.

Precision Grades	Torques (N·m)							
	8	Comparison with Ideal Gears	10	Comparison with Ideal Gears	12	Comparison with Ideal Gears	14	Comparison with Ideal Gears
Ideal	937.19	-	1063.94	-	1166.77	-	1260.01	-
2	1270.99	35.62%	1409.09	32.44%	1527.70	30.93%	1627.02	29.13%
4	1635.07	74.47%	1775.86	66.91%	1902.75	63.08%	2015.56	59.96%
6	2294.29	144.81%	2485.38	133.60%	2627.90	125.23%	2770.99	119.91%

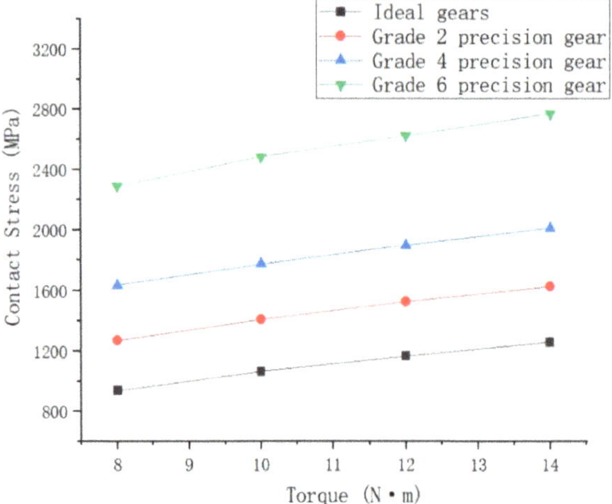

Figure 15. The results of the maximum TSCS under different torques.

3.2. TSCS Analysis of Gear with ME

In this paper, the ME is divided into two directions, and the ME in all directions can be decomposed into these two directions. As shown in Figure 16, point A and point C are the absolute center points of the two engaged gears. When the two gears are in the ideal engaged state, a 3D rectangular coordinate system is established with point A as the origin, where the AC direction is the X axis direction, the AB direction is the Y axis direction (which is the axial direction of gear 1), the Z axis direction is the radial direction of gear 1, plane V is the plane composed of the Y axis and Z axis, and plane S is the plane composed of the X axis and Y axis. Then, the first kind of ME is the rotation error around the Z axis in the S plane, and the amount of meshing error is $<BAB_2$, which is $<a$, and the second kind of ME is the rotation error around the X axis in the V plane, and the amount of ME is $<BAB_1$, which is $<b$.

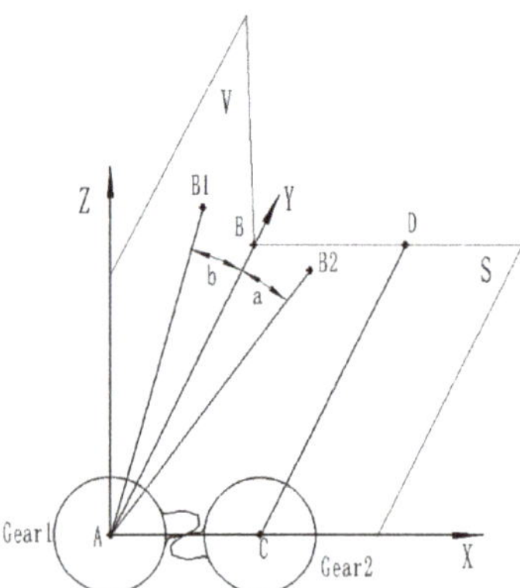

Figure 16. Definitions of meshing errors (ME).

According to the actual measurement data from the gear factory, most of the range of the $<a$ value is 0.1–0.7°, and most of the range of the $<b$ value is 0.02–0.15°. Taking the counterclockwise rotation direction as the positive direction, selecting 0.2°, 0.4° and 0.6° for $<a$ respectively, and conducting the gear FEM contact analysis under the applied torque of 10 N·m, the analysis results are shown in

Figure 17. Then, selecting −0.04°, −0.08° and −0.12° for <*b*, respectively, and conducting the gear FEM contact analysis under the applied torque of 10 N·m, the analysis results are shown in Figure 18.

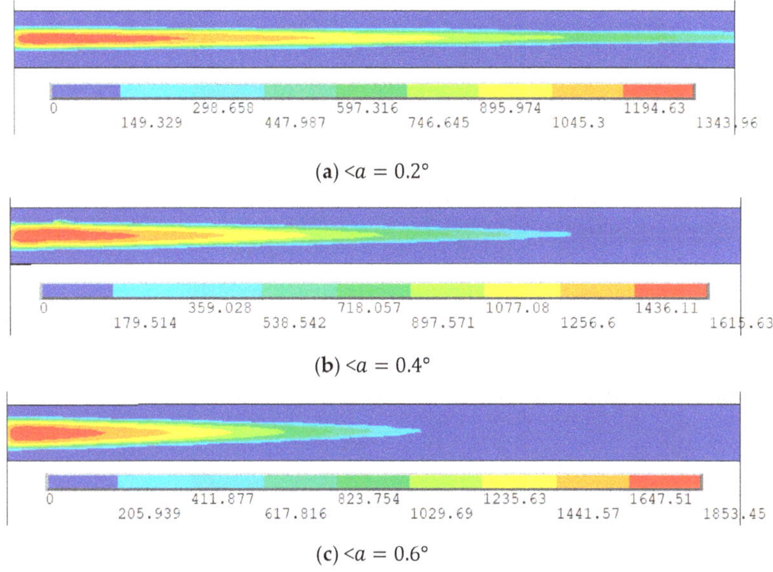

Figure 17. The TSCS nephograms of <*a* = 0.2°, <*a* = 0.4° and <*a* = 0.6°.

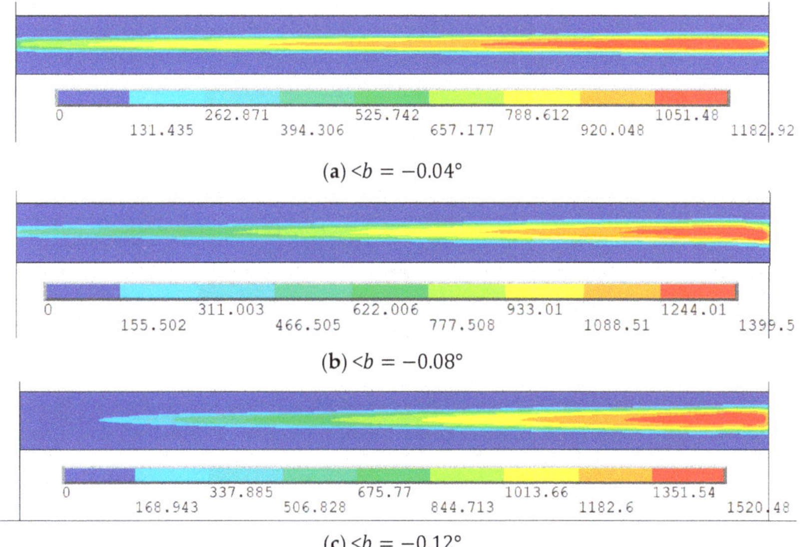

Figure 18. The TSCS nephograms of <*b* = −0.04°, <*b* = −0.08° and <*b* = −0.12°.

In Figure 17, from Figure 17a to Figure 17c are shown the TSCS nephograms of <*a* = 0.2°, <*a* = 0.4° and <*a* = 0.6°. It can be seen that the TSCS with an ME of <*a* is concentrated on the left side of the tooth surface, and with the increase in <*a* value, the gradient of the local TSCS increases gradually, and the maximum TSCS also increases accordingly.

In a similar situation, as can be seen from Figure 18, from Figure 18a to Figure 18c are the TSCS nephograms of $<b = -0.04°$, $<b = -0.08°$ and $<b = -0.12°$. It can be seen that the TSCS with the ME of $<b$ is concentrated on the right side of the tooth surface, and with the increase in $<b$ value, the gradient of the local TSCS increases gradually, and the maximum TSCS also increases accordingly.

In addition, it can be seen from the figures that the TSCS of gears with an ME of $<a$ is concentrated on the left side of the tooth surface, while the TSCS of gears with an ME of $<b$ is concentrated on the right side of the tooth surface. There should be some interaction between the two kinds of ME, which needs further study.

In this paper, the FEM contact analysis is conducted under different torques for gears with different ME, and the results of the maximum TSCS are compared as shown in Table 7 and Figure 19. It can be seen that the influence of ME on the TSCS is very large, and the difference between the maximum TSCS corresponding to the different ME values is large. Within the set range for the ME value, the influence degree of $<a$ is relatively greater. At the same time, it is found that with the increase in applied load, the ratio of the maximum TSCS of each gear a different ME to the ideal gear gradually decreases.

Table 7. The results of the maximum TSCS under different torques. Unit: MPa

ME (°)	Torques (N·m)							
	8	Comparison with Ideal Gears	10	Comparison with Ideal Gears	12	Comparison with Ideal Gears	14	Comparison with Ideal Gears
Ideal	937.19	-	1063.94	-	1166.77	-	1260.01	-
a = 0.2	1266.82	35.17%	1343.96	26.32%	1412.22	21.04%	1499.08	18.97%
a = 0.4	1530.68	63.33%	1615.63	51.85%	1692.28	45.04%	1755.10	39.29%
a = 0.6	1769.67	88.83%	1853.45	74.21%	1949.32	67.07%	2049.15	62.63%
b = −0.04	1081.01	15.35%	1182.92	11.18%	1233.93	5.76%	1314.95	4.36%
b = −0.08	1283.69	36.97%	1399.51	31.54%	1489.06	27.62%	1518.91	24.54%
b = −0.12	1431.54	52.75%	1520.48	42.91%	1588.72	36.16%	1664.76	32.12%

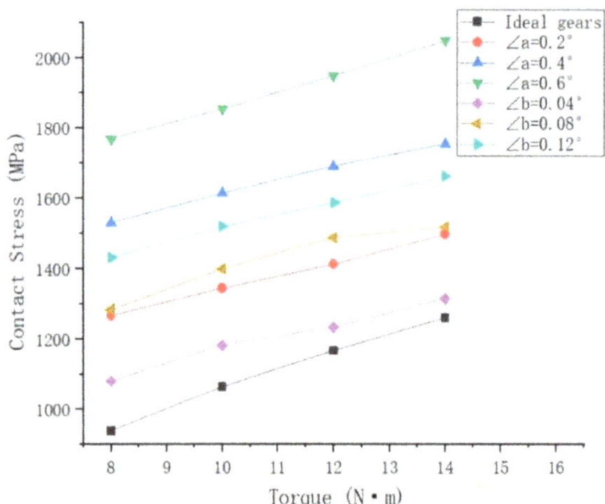

Figure 19. The results of the maximum TSCS under different torques.

3.3. TSCS Analysis of Gear with LCM

The purpose of the LCM is to ensure that when the gears are at the maximum inclination, the teeth engaged with each other do not have an edge point contact (tangent but not intersecting), so as to reduce edge contact.

It should be admitted that there is a certain load concentration phenomenon on the tooth surface with LCM, so the LCM should meet the most basic requirements as follows: first, when the gear tooth has the maximum inclination, the gear tooth shall not have the edge embedding phenomenon; second, when the gear tooth does not incline or has a certain inclination, the gear tooth shall have the minimum load concentration, that is, the lead crowning radius should be the largest (the LCM quantity is the smallest).

According to the ISO standard [4], the LCM mainly considers the ME of the original gear, and the formula is as follows:

$$C = 0.5 \times F_{\beta xcv} \tag{8}$$

where $F_{\beta xcv}$ is the original equivalent misalignment for the determination of the crowning height. According to the range of ME set in this paper, the LCM quantities are taken as 3.5 μm, 7 μm and 10.5 μm, respectively, for the FEM contact analysis.

Figure 20 shows the LCM of the tooth surface, where a is the tooth width, $l = 0.5 \times a$, C is the LCM quantity, and the lead crowning radius is:

$$R = \frac{l^2}{2 \times C} \tag{9}$$

Adjust the tooth surface nodes to input the LCM to the FEM model. Then, the calculation formula of the modification quantity of node i is:

$$C_i = C \times \left(\frac{l_i}{l}\right)^2 \tag{10}$$

where l_i is the distance from the node numbered i to the center line of the tooth width.

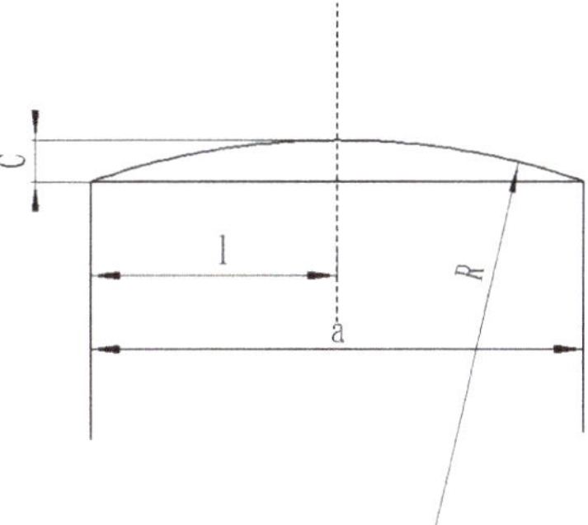

Figure 20. The lead crowning modifications (LCM) of the tooth surface.

In this paper, selecting 3.5 μm, 7 μm and 10.5 μm for the LCM quantities, respectively, and conducting the gear FEM contact analysis under the applied torque of 10 N·m, the analysis results are shown in Figure 21.

In Figure 21, from Figure 21a to Figure 21c are shown the TSCS nephograms of $C = 3.5$ μm, $C = 7$ μm and $C = 10.5$ μm. It can be seen that the TSCS with LCM presents stress the concentration phenomenon in the center of the tooth surface, and with the increase in LCM quantity, the gradient of the local TSCS increases gradually, and the maximum TSCS also increases accordingly.

In this paper, the FEM contact analysis is conducted under different torques for gears with different LCM quantities, and the results of the maximum TSCS are compared as shown in Table 8 and Figure 22. It can be seen that the influence of LCM on the TSCS is very large. At the same time, it is found that with the increase in applied load, the ratio of the maximum TSCS of each gear with different ME to the ideal gear gradually decreases.

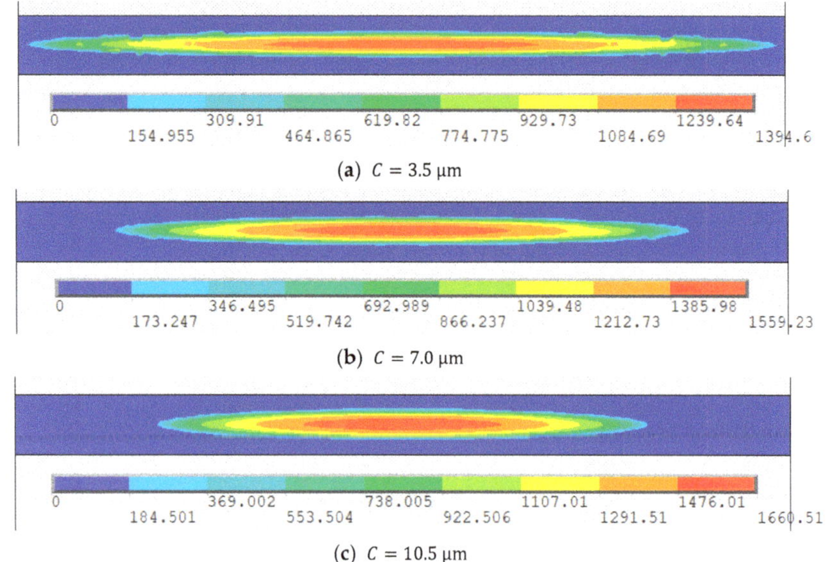

(**a**) $C = 3.5\ \mu m$

(**b**) $C = 7.0\ \mu m$

(**c**) $C = 10.5\ \mu m$

Figure 21. The TSCS nephograms of $C = 3.5\ \mu m$, $C = 7\ \mu m$ and $C = 10.5\ \mu m$.

Table 8. The results of the maximum TSCS under different torques. Unit: MPa.

LCM Quantity (μm)	Torques (N·m)							
	8	Comparison with Ideal Gears	10	Comparison with Ideal Gears	12	Comparison with Ideal Gears	14	Comparison with Ideal Gears
Ideal	937.19	-	1063.94	-	1166.77	-	1260.01	-
2	1285.64	37.18%	1394.60	31.08%	1495.83	28.20%	1587.48	25.99%
4	1444.08	54.09%	1559.23	46.55%	1668.85	43.03%	1754.07	39.21%
6	1541.64	64.50%	1660.51	56.07%	1768.54	51.58%	1851.27	46.93%

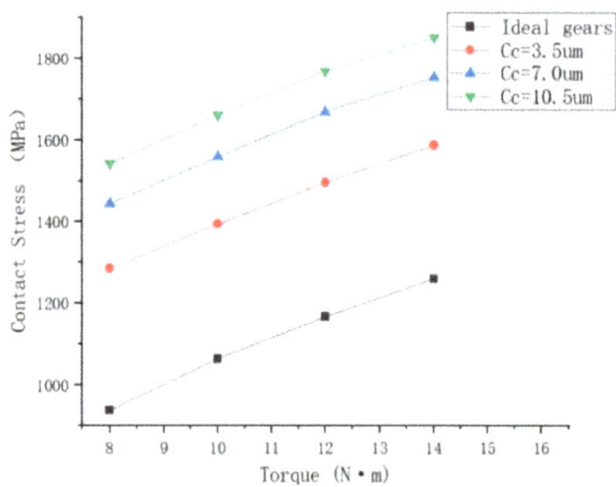

Figure 22. The results of the maximum TSCS under different torques.

The object of the above FEM contact analysis results is a pair of engaged teeth with only LCM and no ME. However, as we all know, in the actual gear engagement process, this must be accompanied by ME and TPD, and the lead crowning machining is also an essential process of gear machining. This is because the LCM can effectively reduce the stress concentration at the edge point caused by the ME and the TPD; therefore, there must be an interaction between these factors, which needs further study.

4. Analysis of the Influence of TPD, ME and LCM on TSCS by TM

According to the content of the third chapter, it can be found that when the TPD, ME and LCM exist separately, the influence on the TSCS is relatively large. However, when the above three factors exist in the gear at the same time, due to the interaction between the factors, the influence degree on the contact stress is not a simple superposition.

This chapter mainly studies the interaction between these three factors, and the influence degree of each factor on the TSCS when the above three factors exist at the same time. As a powerful tool, the Taguchi method can effectively study the above problems, and its steps are shown in Figure 23.

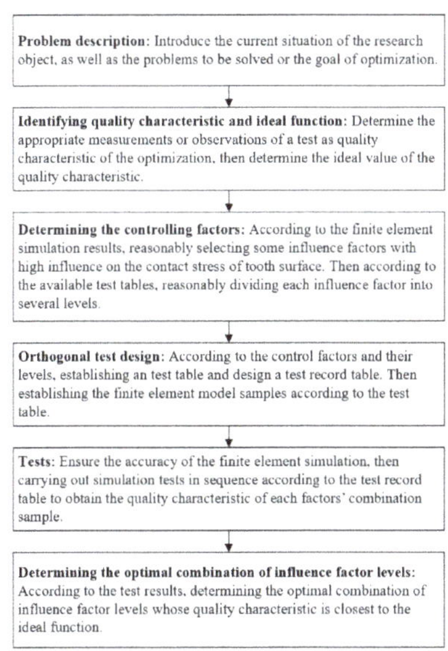

Figure 23. The steps of the Taguchi method (TM).

4.1. Preparations before the Tests

Problem description: When there are three influence factors of TPD, ME and LCM at the same time for a certain type of gear, one must research the influence degree of each factor on the TSCS and the interaction between them, and then determine the optimal combination of factor levels, so as to reduce the TSCS as much as possible.

Quality characteristic and ideal function: Take the maximum TSCS as the quality characteristic. The smaller the maximum TSCS is, the better.

Control factors: The above three factors are the control factors of the Taguchi method. For the TPD, it is divided into three levels (grade 2 precision, grade 4 precision and grade 6 precision). For the ME, $<a$ is divided into three levels (0.2°, 0.4° and 0.6°), and $<b$ is divided into three levels (−0.04°, −0.08° and −0.12°). For the LCM, it is divided into three levels ($C = 3.5$ μm, $C = 7$ μm and $C = 10.5$ μm). The control factor levels are shown in Table 9.

Table 9. The levels of control factors.

Control Factors	TPD	$<a$	$<b$	LCM
Level 1	Grade 2	0.2°	−0.04°	3.5 μm
Level 2	Grade 4	0.4°	−0.08°	7 μm
Level 3	Grade 6	0.6°	−0.12°	10.5 μm

Test design: According to Table 9, the next step is to establish an orthogonal test table to record the test data, and the corresponding FEM simulation model is established according to the orthogonal test table. In this paper, Minitab is used to generate the orthogonal test table. There are two kinds of test tables with four factors and three levels. One has nine groups of tests, and the other has 27 groups of tests. Because we want not only to get the optimal combination of the factor levels, but also to research the interaction between various factors, 27 groups of tests are selected. The test table is shown in Table 10.

Table 10. The orthogonal test table.

Serial Number	Tooth Profile Precision Grade	$<b$ (°)	$<a$ (°)	Lead Crowning (μm)
1	Grade 2	−0.04	0.2	3.5
2	Grade 2	−0.04	0.4	7.0
3	Grade 2	−0.04	0.6	10.5
4	Grade 4	−0.08	0.2	3.5
5	Grade 4	−0.08	0.4	7.0
6	Grade 4	−0.08	0.6	10.5
7	Grade 6	−0.12	0.2	3.5
8	Grade 6	−0.12	0.4	7.0
9	Grade 6	−0.12	0.6	10.5
10	Grade 4	−0.12	0.2	7.0
11	Grade 4	−0.12	0.4	10.5
12	Grade 4	−0.12	0.6	3.5
13	Grade 6	−0.04	0.2	7.0
14	Grade 6	−0.04	0.4	10.5
15	Grade 6	−0.04	0.6	3.5
16	Grade 2	−0.08	0.2	7.0
17	Grade 2	−0.08	0.4	10.5
18	Grade 2	−0.08	0.6	3.5
19	Grade 6	−0.08	0.2	10.5
20	Grade 6	−0.08	0.4	3.5
21	Grade 6	−0.08	0.6	7.0
22	Grade 2	−0.12	0.2	10.5
23	Grade 2	−0.12	0.4	3.5
24	Grade 2	−0.12	0.6	7.0
25	Grade 4	−0.04	0.2	10.5
26	Grade 4	−0.04	0.4	3.5
27	Grade 4	−0.04	0.6	7.0

4.2. Analysis of FEM Simulation Results

According to Table 9, 27 FEM simulation models with different factor level combinations are established and solved.

Figure 24 is the TSCS nephogram of the first FEM model in Table 10, and compared with the TSCS nephogram with only one single influence factor, it has the characteristics of irregularity of TPD, an edge stress concentration of ME and a central stress concentration of LCM. Table 11 shows the results of 27 FEM simulation tests.

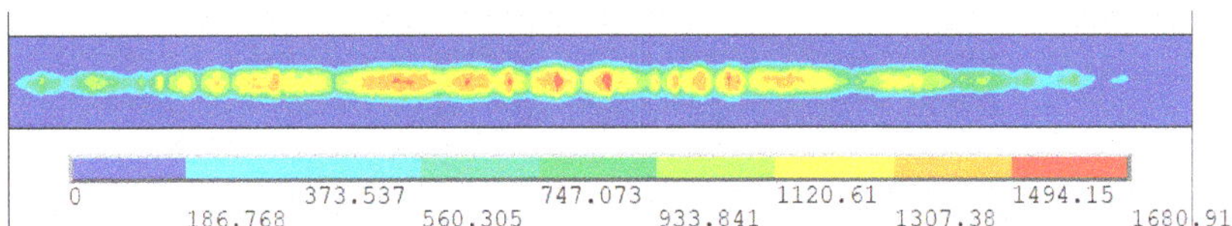

Figure 24. The TSCS nephogram of the first FEM model.

Table 11. The maximum TSCS of 27 FEM simulation tests.

Serial Number	Maximum TSCS (MPa)	Serial Number	Maximum TSCS (MPa)	Serial Number	Maximum TSCS (MPa)
1	1680.91	10	2272.12	19	3888.53
2	1854.25	11	2459.15	20	3070.89
3	1931.67	12	2135.32	21	3636.82
4	2118.22	13	3719.14	22	1962.06
5	2322.13	14	3819.85	23	1683.50
6	2420.97	15	3229.20	24	1850.24
7	3102.55	16	1907.33	25	2454.60
8	3753.72	17	1999.51	26	2177.90
9	3861.61	18	1806.02	27	2297.35

Minitab (17.1.0, Minitab, LLC, State College, PA, USA) was used to analyze the test results. The Minitab response tables for signal to noise (SN) ratios and means are shown in Table 12. From the table, whether it is the SN ratios response table or the mean response table, it can be seen that the TPD has the greatest influence on the TSCS, followed by the LCM; the $<b$ and $<a$ are relatively much smaller.

Table 12. The Minitab response tables for signal to noise (SN) ratios and means.

Level	Tooth Profile Accuracy		$<b$ (°)		$<a$ (°)		Lead Crowning (μm)	
	SN Ratios	Means	SN Ratios	Means	SN Ratios	Means	SN Ratios	Means
1	−65.34	1853	−67.85	2574	−67.83	2567	−67.09	2334
2	−67.20	2295	−67.89	2574	−67.85	2571	−68.01	2624
3	−71.01	3565	−67.81	2564	−67.87	2574	−68.45	2755
Delta	5.66	1712	0.08	10	0.04	7	1.36	421
Rank	1		3		4		2	

The main effects plots for SN ratios and means are shown in Figure 25. From the Figure 25a,b, within the set value range, it can be seen that the influence degree of TPD on the TSCS is much greater than that of LCM, and the influence degree of LCM is much greater than that of ME. The influence degree of the two kind of ME is very small, and the difference is not big. In addition, it can be seen that when the precision grade changes from 4 to 6, it has a great influence on the TSCS; when changing from 2 to 4, the influence degree is significantly reduced. Similarly, for the LCM, when the LCM quantity changes from 3.5 μm to 7 μm, the influence degree is also slightly larger.

4.3. The Interaction between TPD, ME and LCM

The interaction between different factors can be obtained through Minitab. Figure 26 shows the interaction plot for means between the TPD and other factors. The nonparallel lines in the interaction plot indicate the interaction degree between the TPD and other factors. If the three lines are completely parallel, it means that there is no interaction between them; on the contrary, the more nonparallel the three lines, the greater the interaction degree between them.

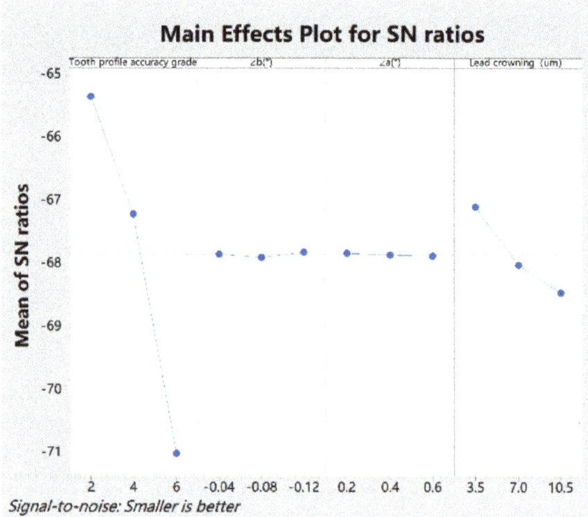

(**a**) The main effects plots for signal to noise (SN) ratios.

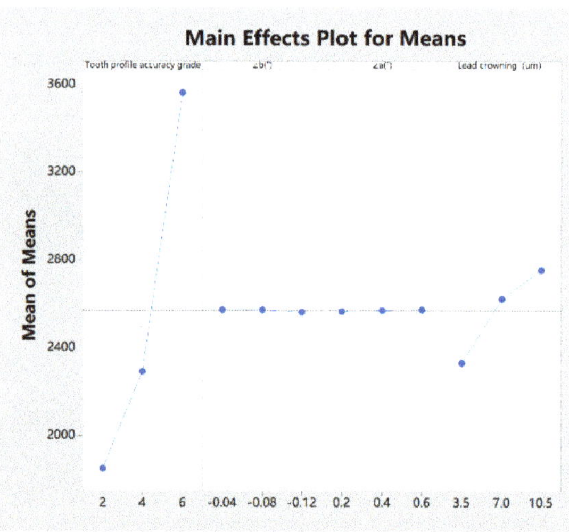

(**b**) The main effects plots for means.

Figure 25. The main effects plots for signal to noise (SN) ratios and means.

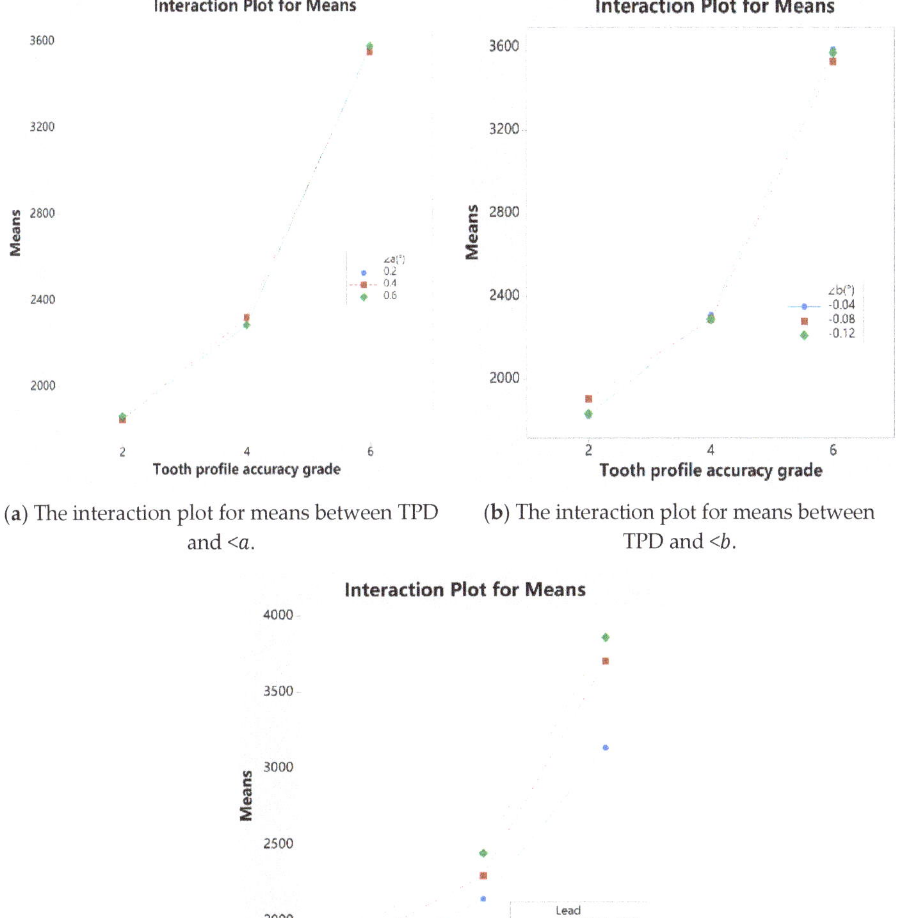

(**a**) The interaction plot for means between TPD and $\angle a$.

(**b**) The interaction plot for means between TPD and $\angle b$.

(**c**) The interaction plot for means between TPD and LCM.

Figure 26. The interaction plot for means between the TPD and other factors.

As can be seen from Figure 26a,b, there is basically no interaction between the TPD and the ME, and from Figure 26c, there is a certain interaction between the TPD and the LCM, but the interaction degree is not significant.

Figure 27a,b show the interaction plots for means between the LCM and the ME, and Figure 27c shows the interaction plot for means between $\angle a$ and $\angle b$. From Figure 27a, it can be seen that there is an obvious interaction between the LCM and the $\angle a$; in particular, when the $\angle a$ changes from 0.4° to 0.6°, the interaction degree is more significant. From Figure 27b, it can be seen that there is also an obvious interaction between the LCM and the $\angle b$. Therefore, there is an obvious interaction between the LCM and the overall ME. From Figure 27c, it can be seen that the interaction degree between meshing errors $\angle a$ and $\angle b$ is very significant, even though the influence degree of these two kinds of ME on the TSCS is very small.

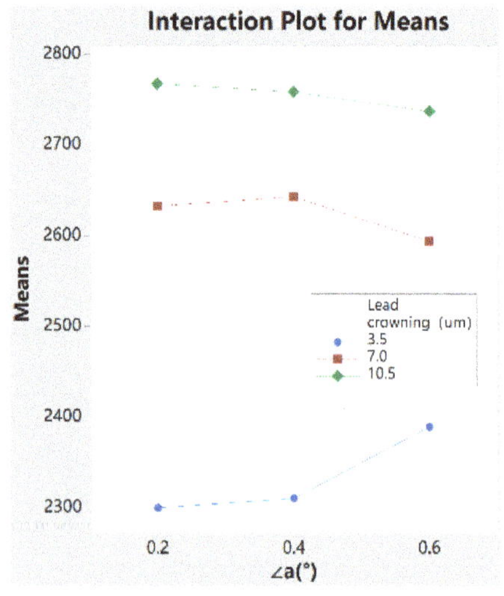

(a) The interaction plot for means between $\angle a$ and LCM.

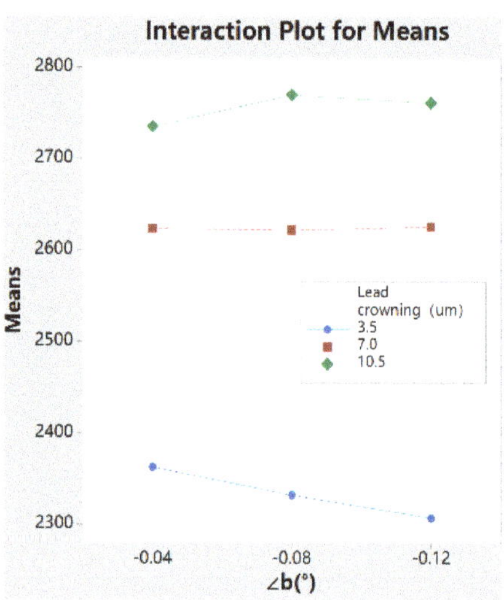

(b) The interaction plot for means between $\angle b$ and LCM.

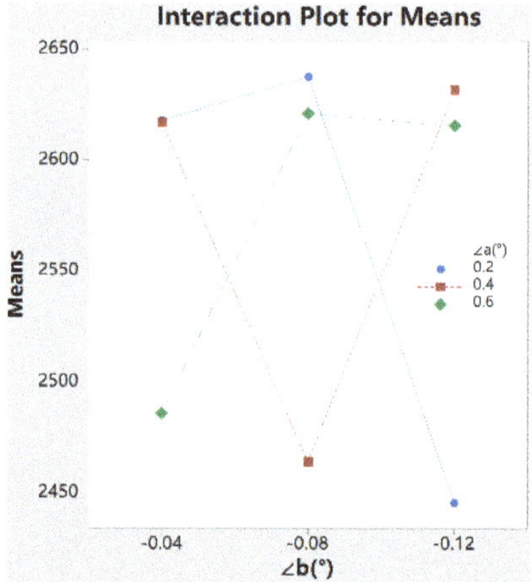

(c) The interaction plot for means between $\angle a$ and $\angle b$.

Figure 27. The interaction plot for means.

It can be known from the third chapter that when the $\angle a$ and $\angle b$ exist alone, they have a great influence on the TSCS. However, when all the influence factors exist, because of the obvious interaction between them, the influence degree of the $\angle a$ and $\angle b$ on the TSCS becomes very small.

In addition, in order to further verify the importance of LCM for the engaged gears with ME, two FEM simulation models are established. The two models have the same precision grade of TPD (grade 2), $\angle a$ (0.2°) and $\angle b$ (−0.04°), but the first model does not have the LCM, and the second model has an LCM of 3.5 um. The TSCS analysis of the two models is carried out, and Figure 28a is the stress nephogram of the first model, while Figure 28b is the stress nephogram of the second model. Through comparison, it can be found that the LCM can effectively alleviate the edge stress concentration caused by ME, and the maximum TSCS of the model with LCM is significantly smaller than that without LCM. This also demonstrates that there is an obvious interaction between LCM and ME.

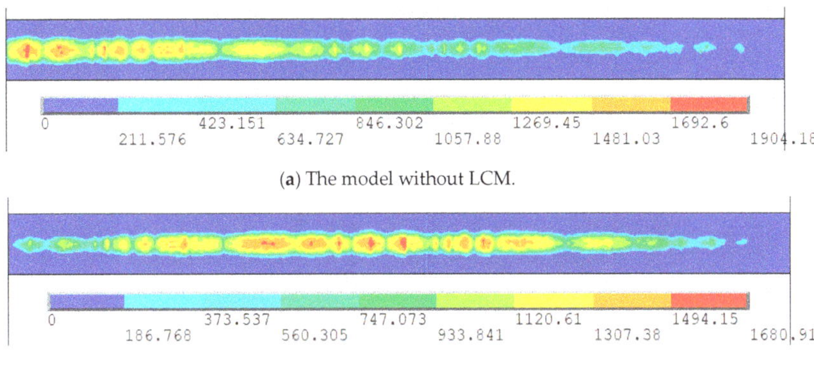

(**a**) The model without LCM.

(**b**) The model with LCM.

Figure 28. The TSCS nephogram.

4.4. Determining the Optimal Combination of Influence Factor Levels

The maximum TSCS of each combination of factor levels can also be predicted by the Minitab. As shown in Figure 29, the approximate maximum TSCS corresponding to the combination of different factor levels can be obtained by inputting the level of each factor.

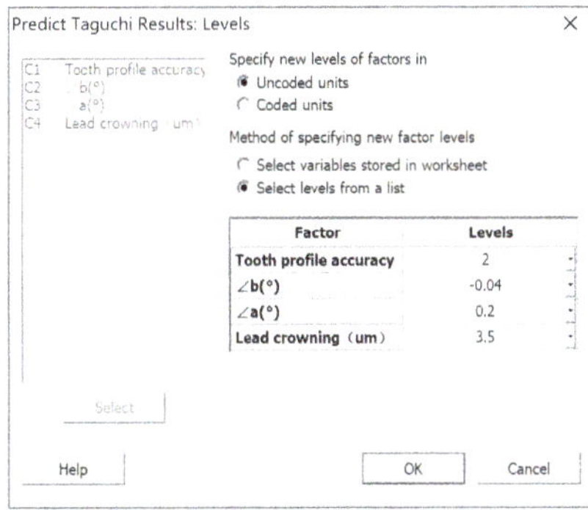

Figure 29. The Taguchi results are predicted by Minitab.

It can be seen from Figure 25 that the TPD grade is 2 and the LCM quantity is 3.5 µm in the optimal combination of factor levels. Then, the prediction results of partial combinations are shown in Table 13. It can be seen from Table 13 that when the TPD grade is 2 and the LCM quantity is 3.5 µm, the ME has little effect on the maximum TSCS, and the maximum TSCS range is 1605.57–1622.67 MPa. Therefore, it is sufficient to control the $\angle a$ and $\angle b$ of ME within the specified range. However, in conclusion, it is determined that the optimal combination of factor levels is grade 2 for the TPD, −0.12° for the $\angle b$, 0.2° for the $\angle a$, and 3.5 µm for the LCM quantity.

Table 13. The prediction results of different combinations.

Table	$<b$ (°)	$<a$ (°)	LCM (μm)	TSCS (MPa)
		0.2		1614.97
	−0.04	0.4		1618.91
		0.6		1622.06
		0.2		1615.59
2	−0.08	0.4	3.5	1619.53
		0.6		1622.67
		0.2		1605.57
	−0.12	0.4		1609.51
		0.6		1612.66

4.5. Comparative Analysis of Different Combinations of Factor Levels

According to the original process requirements of precision gear manufacturing in a gear factory, the TPD, $<a$, $<b$ and LCM are required to be controlled at grade 4, 0.4°, −0.08° and 3.5 μm, respectively, and the maximum TSCS of the original combination of factor levels is 2062.00 MPa, which is 28.43% larger than the 1605.57 MPa of the optimal combination.

In order to better show the influence of the TSCS on the gear fatigue life, according to the contact fatigue SN curve (Figure 30) and the corresponding curve equation (Equation (11)) provided by the gear factory, the gear fatigue life corresponding to different TSCS can be obtained.

$$\lg\sigma = -0.0908\lg N + 3.8767 \tag{11}$$

According to Equation (11), when the $\sigma = 1605.57$ MPa, the corresponding fatigue life $N = 2.46 \times 10^7$, while when the maximum TSCS is 2062 MPa, the corresponding fatigue life $N = 1.5 \times 10^6$. It can be found that the gear contact fatigue life of the optimal combination of factor levels is much longer than that of the original combination.

In addition, when the minimum fatigue life of the gear is required to be 1×10^7, the maximum acceptable TSCS is 1743 MPa, as calculated by Equation (11). Therefore, only when the TPD is grade 2 and the LCM is 3.5 μm can the fatigue life meet the requirements. It also shows that the contact fatigue performance of the optimal combination of factor levels is better than that of the original combination.

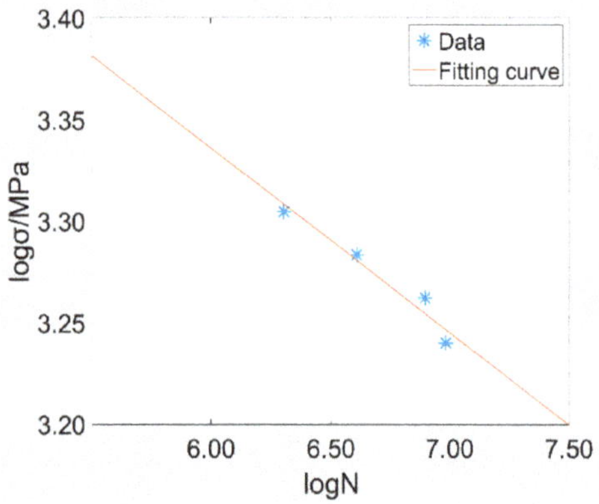

Figure 30. The contact fatigue SN curve.

5. Conclusions

Based on the 3D-FEM and TM, this paper has analyzed the TSCS of spur gears with three different influence factors: TPD, ME and LCM. In particular, it has analyzed the interactions between PD, ME and LCM, and their influence degrees on the TSCS. Throughout the full paper, there are several conclusions that need to be summarized, as follows:

(1) In this paper, a 3D-FEM model of one pair of engaged teeth has been modeled and the mesh of the contact area has been refined by FEM software. In the model, the refined area mesh and the non-refined area mesh were connected by Multi-point constraint (MPC). At the same time, in order to save the time of the FEM solution on the premise of ensuring the accuracy of model analysis, the reasonable size of the refined area has been researched and confirmed;

(2) In this paper, the FEM contact models of gears have been established and solved according to the three influence factors of TPD, ME and LCM, respectively. It is found that when there is only one single influence factor, the influence factor has a great influence on the TSCS. Compared with the ideal gear, the maximum TSCS of the gear with TPD can reach 2.45 times of that of the ideal gear, the maximum TSCS of the gear with ME can reach 1.88 times of that of the ideal gear, and the maximum TSCS of the gear with LCM can reach 1.65 times of that of the ideal gear;

(3) In this paper, the Taguchi method has been used to research the influence degree of each factor on the TSCS when three factors exist at the same time. It is found that the TPD has the greatest influence on the TSCS, as the mean difference in TSCS between different grades of TPD can reach 1712 MPa. This is followed by the LCM, as the mean difference in TSCS between different LCM values can reach 421 MPa. The influence degree of ME is very limited, as the mean differences in TSCS value between different $<a$ and $<b$ are only 7 MPa and 10 MPa, respectively;

(4) In this paper, the interactions between the influence factors have been researched, and it is found that the interactions between the TPD and other factors are not obvious, while the interactions between the LCM, $<a$ and $<b$ are very obvious, especially the interaction between the two kinds of ME. In addition, it has been verified that the LCM can effectively alleviate the phenomenon of the edge stress concentration of TSCS caused by ME. From Figure 28, it can be found that the maximum TSCS value of the gear with LCM is 11.73% less than that of the gear without LCM;

(5) According to the type of gear researched in this paper, through the TM, the optimal combination of control factor levels has been determined, as follows: grade 2 for the TPD, $-0.12°$ for the $<b$, $0.2°$ for the $<a$, and 3.5 μm for the LCM quantity. The gear contact fatigue life of the optimal combination of factor levels is much longer than that of the original combination;

(6) For other types of gears and different influence factors, the research method and analysis process of this paper have certain reference value.

Author Contributions: Writing—original draft, Q.L.; Writing—review and editing, L.X. All authors have read and agreed to the published version of the manuscript.

Funding: This research was funded by the "Research on Reliability Theory and Method of Total Fatigue Life for Large and Complex Mechanical Structures" of Natural Science Foundation of China and Liaoning Province, grant number U1708255.

Conflicts of Interest: The authors declare no conflict of interest.

Abbreviations

3D	Three-dimensional
FEM	Finite element method
LCM	Lead crowning modifications
ME	Meshing errors
MPC	Multi-point constraint
TM	Taguchi method
TPD	Tooth profile deviations
TSCS	Tooth surface contact stress

References

1. Wei, J.; Lv, C.; Sun, W.; Li, X.; Wang, Y. A study on optimum design method of gear transmission system for wind turbine. *Int. J. Precis. Eng. Manufact.* **2013**, *14*, 767–778. [CrossRef]
2. Li, M.; Xie, L.; Ding, L. Load sharing analysis and reliability prediction for planetary gear train of helicopter. *Mechan. Mach. Theor.* **2017**, *115*, 97–113. [CrossRef]
3. ISO 6336-2: 2019. Calculation of Load Capacity of Spur and Helical Gears–Part. 2: Calculation of Surface Durability (Pitting). Available online: https://www.iso.org/standard/63821.html (accessed on 6 September 2020).
4. ISO 6336-1. Calculation of Load Capacity of Spur and Helical Gears–Part. 1: Basic Principles, Introduction and General Influence Factors. Available online: https://www.iso.org/standard/63819.html (accessed on 6 September 2020).
5. Li, S. Finite element analyses for contact strength and bending strength of a pair of spur gears with machining errors, assembly errors and tooth modifications. *Mech. Mach. Theory* **2007**, *42*, 88–114. [CrossRef]
6. Sennba, M. *Gear Errors and Strength*; The Nikkan Kogyo Shimbun Press: Tokyo, Japan, 1974; pp. 70–148. (In Japanese)
7. Ottewill, J.; Neild, S.A.; Wilson, R. An investigation into the effect of tooth profile errors on gear rattle. *J. Sound Vib.* **2010**, *329*, 3495–3506. [CrossRef]
8. Ma, P.; Botman, M. Load Sharing in a Planetary Gear Stage in the Presence of Gear Errors and Misalignment. *J. Mechan. Design* **1985**, *107*, 4–10. [CrossRef]
9. Bodas, A.; Kahraman, A. Influence of carrier and gear manufacturing errors on the static planet load sharing behavior of planetary gear sets (gear unit design and applications). *Japan Soc. Mechan. Engr.* **2017**. [CrossRef]
10. Miyoshi, Y.; Tobisawa, K.; Saiki, K. Composite analysis method of tooth contact load distribution of helical gear. In Proceedings of the ASME International Design Engrring Technical Conferences/Computers and Information in Engrring Conference, New York, NY, USA, 4–7 September 2007; pp. 173–180.
11. Wang, Q.B.; Hu, P.; Zhang, Y.; Wang, Y.; Pang, X.; Tong, C. A model to determine mesh characteristics in a gear pair with tooth profile error. *Adv. Mechan. Engr.* **2014**, *6*, 1–10. [CrossRef]
12. Lin, T.; He, Z. Analytical method for coupled transmission error of helical gear system with machining errors, assembly errors and tooth modifications. *Mech. Syst. Signal. Process.* **2017**, *91*, 167–182. [CrossRef]
13. Li, Q.; Xie, L.; Song, J.; Li, H.; Xu, G. Research Methods and Applications of Gear Manufacturing Process Optimization. *Mathem. Probl. Eng.* **2019**, *2019*, 1–17. [CrossRef]
14. Yu, B.S.; Liu, Y.Y. Improvement in phase purity and yield of hydrothermally synthesized smectite using Taguchi method. *Appl. Clay Sci.* **2018**, *161*, 103–109. [CrossRef]
15. Sun, B.; Xiao, J.; Li, Z.; Ma, B.; Zhang, L.; Huang, Y.; Bai, L. An analysis of soil detachment capacity under freeze-thaw conditions using the Taguchi method. *Catena* **2018**, *162*, 100–107. [CrossRef]
16. Mojtaba, F.; Abbas, N.; Sima, Z.; Mohammad, R.H. Optimization of H-shaped thin-walled energy absorber by Taguchi method and a new theoretical estimation for its energy absorption. *Thin Wall. Struct.* **2018**, *131*, 33–44.
17. Ji, D.; Wei, Z.; Mazzoni, S.; Mengarelli, M.; Rajoo, S.; Zhao, J.; Pou, J.; Romagnoli, A. Thermoelectric generation for waste heat recovery: Application of a system level design optimization approach via Taguchi method. *Energy Convers. Manag.* **2018**, *172*, 507–516. [CrossRef]
18. Shuting, L. Effects of misalignment error, tooth modifications and transmitted torque on tooth engagements of a pair of spur gears. *Mechan. Machine Theory* **2015**, *83*, 125–136.

19. Hertz, H. On the contact of elastic solids. *J. Reine Angew Math.* **1882**, *92*, 156–171.
20. ISO 1328-1. Cylindrical Gears–ISO System of Flank Tolerance Classification–Part. 1: Definitions and Allowable Values of Deviations Relevant to Flanks of Gear Teeth. Available online: https://www.iso.org/obp/ui/#iso:std:iso:1328:-1:ed-2:v1:en (accessed on 6 September 2020).

Publisher's Note: MDPI stays neutral with regard to jurisdictional claims in published maps and institutional affiliations.

Article

Early Crack Propagation in Single Tooth Bending Fatigue: Combination of Finite Element Analysis and Critical-Planes Fatigue Criteria

Franco Concli [1,*], **Lorenzo Maccioni** [1,*], **Lorenzo Fraccaroli** [1] and **Luca Bonaiti** [2]

[1] Faculty of Science and Technology, Free University of Bolzano/Bozen, Piazza Università 1, 39100 Bolzano, Italy; lorenzo.fraccaroli@unibz.it

[2] Department of Mechanical Engineering, Politecnico di Milano, via La Masa 1, 20157 Milano, Italy; luca.bonaiti@polimi.it

* Correspondence: franco.concli@unibz.it (F.C.); lorenzo.maccioni@unibz.it (L.M.)

Abstract: Mechanical components, such as gears, are usually subjected to variable loads that induce multiaxial non-proportional stress states, which in turn can lead to failure due to fatigue. However, the material properties are usually available in the forms of bending or shear fatigue limits. Multiaxial fatigue criteria can be used to bridge the gap between the available data and the actual loading conditions. However, different criteria could lead to different results. The main goal of this paper is to evaluate the accuracy of different criteria applied to real mechanical components. With respect to this, five different criteria based on the critical plane concept (i.e., Findley, Matake, McDiarmid, Papadopoulos, and Susmel) have been investigated. These criteria were selected because they not only assess the level of damage, but also predict the direction of crack propagation just after nucleation. Therefore, measurements (crack position and direction) on different fractured gear samples tested via Single Tooth Bending Fatigue (STBF) tests on two gear geometries were used as reference. The STBF configuration was numerically simulated via Finite Elements (FE) analyses. The results of FE were elaborated based on the above-mentioned criteria. The numerical results were compared with the experimental ones. The result of the comparison showed that all the fatigue criteria agree in identifying the most critical point. The Findley and Papadopulus criteria proved to be the most accurate in estimating the level of damage. The Susmel criterion turns out to be the most conservative one. With respect to the identification of the direction of early propagation of the crack, the Findley criterion revealed the most appropriate.

Keywords: gears; Single Tooth Bending Fatigue; STBF; Finite Element Model; FEM; material characterization; multiaxial fatigue; critical plane

Citation: Concli, F.; Maccioni, L.; Fraccaroli, L.; Bonaiti, L. Early Crack Propagation in Single Tooth Bending Fatigue: Combination of Finite Element Analysis and Critical-Planes Fatigue Criteria. *Metals* **2021**, *11*, 1871. https://doi.org/10.3390/met11111871

Academic Editors: Filippo Berto, Ricardo Branco and Shengchuan Wu

Received: 28 October 2021
Accepted: 17 November 2021
Published: 21 November 2021

1. Introduction

In mechanical systems, gears are widely used components to transmit torque and motion (i.e., mechanical power) between non-coaxial shafts [1]. Due to their working principle (i.e., meshing of conjugate profiles), teeth are subject to various damage mechanisms that can lead to the failure of the entire mechanical system [2,3]. Wear, scuffing, and (micro)pitting in the teeth flank are just a few examples of failure modes that, in turn, can be attributable to high contact pressures and/or insufficient lubrication [4–8]. However, Tooth (root) Bending Fatigue (TBF) is the most dangerous one [9,10].

TBF leads to the nucleation and propagation of a crack in the Tooth Root Radius (ρ_{fP}) due to the varying stress induced by tooth bending during meshing [11,12]. Therefore, a fundamental aspect to be considered while designing gears is the capability to withstand cyclic bending loads [13]. With this respect, different standards support the gear design to avoid TBF failures, e.g., ISO 6336-3 [14,15] and ANSI/AGMA [16]. The above-mentioned standards support gear design through the determination of Tooth Bending Strength (TBS).

According to the Method B of ISO 6336-3 [15], the maximum stress σ_F at the ρ_{fP} due to pure bending has to not exceed the permissible bending stress σ_{FP} that, in turn, is proportional to the material strength σ_{Flim}, i.e., a material property usually determined through Single Tooth Bending Fatigue (STBF) tests [17–20].

In STBF tests, teeth belonging to a gear sample made of the material to be characterized are loaded with two pulsating, competing, parallel, and discordant forces applied through two anvils (having parallel faces) mounted on a universal (fatigue) testing machine [21,22]. Exploiting the Wildhaber property [1], these forces are applied perpendicularly to two tooth flanks, resulting in tangent to the base circle [23]. STBF tests are interrupted if a tooth fails or if it withstands the run-out condition $(10^6–10^8)$ cycles [24–28]. The statistical elaboration of failures and run-outs (the load levels are defined with the staircase approach [29,30]) leads to the determination of the load-carrying capacity [20]. The methods for translating this load into stresses acting at the ρ_{fP} (i.e., σ_{Flim}) can be different: (1) through the reverse application of the standard (e.g., [31–34]), (2) by means of experimental measures (e.g., exploiting strain gauges [35]), (3) via Finite Element (FE) simulations (e.g., [36–39]).

With respect to the FE simulations, on the one hand, they allow us to obtain relevant information on the principal stresses in the ρ_{fP}, for each loading condition. On the other hand, results of FE simulation have to be further elaborated to estimate the fatigue behavior of a specific gear design [40,41]. In other words, numerical simulations of STBF tests and further data elaboration based on fatigue criteria using material data obtained via standard tests (i.e., torsion, bending and traction quasi-static and fatigue tests on standard specimens) seems to be a valuable alternative to long experimental campaigns.

In recent studies, the authors pointed out that the FE simulation results of the STBF configuration can be analyzed and elaborated via different fatigue criteria based on the critical plane approach [41–43]. These allow for evaluating the criticality of each point along the ρ_{fP}. Moreover, it permits to individuate the potential propagation direction of the crack after nucleation. Nevertheless, it has been observed that different fatigue criteria could lead to different results in terms of TBS and/or crack propagation direction [42].

The goal of the present paper is to evaluate the most appropriate fatigue criteria for characterizing the fatigue behavior in terms of the individuation of the nucleation point and the determination of the direction of early propagation of the crack in real mechanical components characterized by non-proportional multiaxial states of stress. This stress state, i.e., any state of time varying stress where the orientation of the principal axes changes with respect to a reference system integral with the studied component, can be found in gears [38]. In this respect, STBF tests described in [35,38–40] have been numerically reproduced and the FE results have been analyzed through different fatigue criteria based on the critical plane, i.e., Findley [44], Matake [45], McDiarmid [46], Papadopoulos [47], and Susmel et al. [48]. The outcomes of the elaboration have been compared with the cracks observed in the above-mentioned experimental campaigns [35,38–40].

The present paper is organized as follows. In Section 2, the mathematical elaboration of a generic time-dependent stress tensor $\overline{\overline{\sigma}}(t)$ according to different fatigue criteria is presented. In Section 3, FE modeling, numerical data processing, and experimental data acquisition are shown. Comparison of numerical and experimental results are presented in Section 4. Discussions and conclusion can be found in Section 5.

2. Background: Mathematical Modeling of Fatigue Criteria Based on the Critical Plane

In the present section, the mathematical modeling is presented of the main fatigue criteria based on critical plane (i.e., Findley [44], Matake [45], McDiarmid [46], Papadopoulos [47], and Susmel et al. [48]). Each fatigue criterion starts from the time-dependent stress tensor $\overline{\overline{\sigma}}(t)$ (Equation (1)) referred to the point whose fatigue behavior has to be evaluated.

$$\overline{\overline{\sigma}}(t) = \begin{bmatrix} \sigma_{xx}(t) & \tau_{xy}(t) & \tau_{xz}(t) \\ \tau_{yx}(t) & \sigma_{yy}(t) & \tau_{yz}(t) \\ \tau_{zx}(t) & \tau_{zy}(t) & \sigma_{zz}(t) \end{bmatrix} \tag{1}$$

More specifically, a generic plane (including the point to be evaluated) can be defined by means of its normal vector n that, in turn, can be defined according with its spherical coordinates (φ_n, ϑ_n) with respect to a generic reference system (Figure 1).

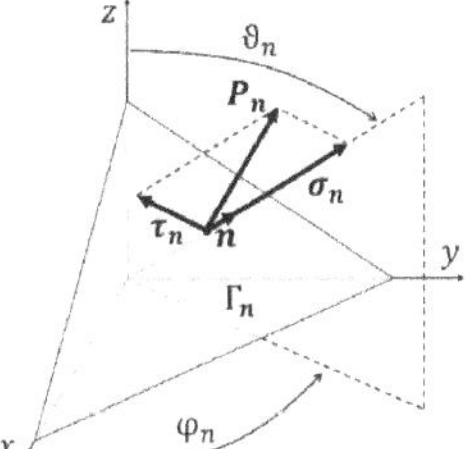

Figure 1. Components of $P_n(\varphi_n, \vartheta_n, t)$ on the plane $n(\varphi_n, \vartheta_n)$.

According to Equation (2), it is possible to calculate the stress vector P_n acting on the afore-mentioned plane; the vector can, in turn, be decomposed into a normal component σ_n and into a tangential component τ_n (Figure 1).

$$P_n(\varphi_n, \vartheta_n, t) = \overline{\overline{\sigma}}(t)\, n(\varphi_n, \vartheta_n) \tag{2}$$

On the one hand, σ_n (which can be calculated through Equation (3)) presents a fixed direction and a time-dependent modulus. On the other hand, τ_n has a time-dependent modulus and direction. Therefore, τ_n has to be further decomposed into its components along the u and v directions (Figure 2). The unitary vectors n, u, v are defined as in Equation (4). In Equation (5), τ_n is defined.

$$\sigma_n(\varphi_n, \vartheta_n, t) = n^T(\varphi_n, \vartheta_n)\overline{\overline{\sigma}}(t)\, n(\varphi_n, \vartheta_n) \tag{3}$$

$$u(\varphi_n, \vartheta_n) = \begin{bmatrix} -\sin\vartheta_n \\ \cos\varphi_n \\ 0 \end{bmatrix} v(\varphi_n, \vartheta_n) = \begin{bmatrix} -\cos\vartheta_n\cos\varphi_n \\ -\cos\vartheta_n\sin\varphi_n \\ \sin\theta_n \end{bmatrix}; \; n(\varphi_n, \vartheta_n) = \begin{bmatrix} \sin\vartheta_n\cos\varphi_n \\ \sin\vartheta_n\sin\varphi_n \\ \cos\vartheta_n \end{bmatrix} \tag{4}$$

$$\tau_n(\varphi_n, \vartheta_n, t) = u^T(\varphi_n, \vartheta_n)\overline{\overline{\sigma}}(t)u(\varphi_n, \vartheta_n) + v^T(\varphi_n, \vartheta_n)\overline{\overline{\sigma}}(t)v(\varphi_n, \vartheta_n) \tag{5}$$

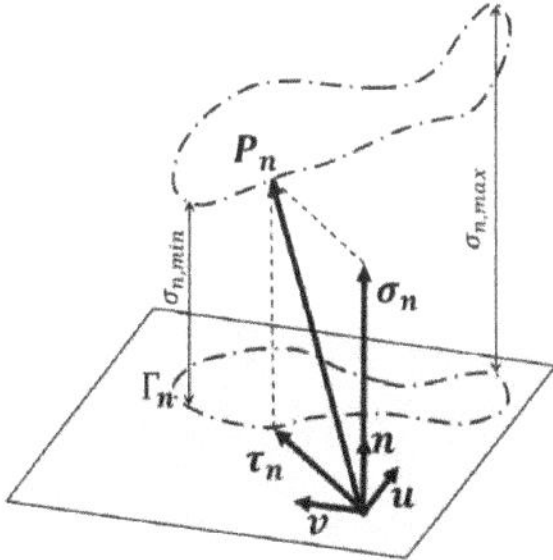

Figure 2. u and v on the plane $n(\varphi_n, \vartheta_n)$ and definition of the curve Γ_n.

It is worth noting that, for periodic stresses (having a period T), the point of the arrow of the vector P_n describes a closed tridimensional curve. Consequently, τ_n describes a closed curve in the plane (Figure 2). In the Figure, this curve is indicated as Γ_n. On the one hand, the normal stress σ_n ranges from a minimum $\sigma_{n,min}$ to a maximum $\sigma_{n,max}$ value (Figure 2). Therefore, it is possible to define the value of the alternating stress (acting on the plane having normal n) as $\sigma_{n,a}$ defined according to Equation (6). On the other hand, to define the value of alternate tangential stress $\tau_{n,a}$ (acting on the plane having normal n), literature reports different methods. The most diffused one is the Minimum Circumscribed

Circle (MCC) (Equation (7)) [49]. Considering that the curve Γ_n is representative of the tangential stresses acting on the studied plane during the entire loading cycle, the MCC method suggests determining $\tau_{n,a}$ as the radius of the smallest circle that can entirely contain the curve Γ_n (Figure 3).

$$\sigma_{n,a} = \max_T\{\sigma_n(t)\} - \min_T\{\sigma_n(t)\} = \sigma_{n,max} - \sigma_{n,min} \tag{6}$$

$$\tau_{n,a} = \underset{T}{MCC}\{\tau_n(t)\} \tag{7}$$

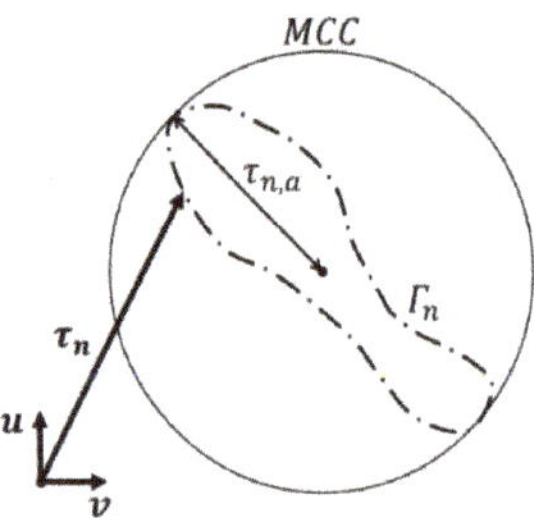

Figure 3. Minimum Circumscribed Circle (MCC) method.

By varying the spherical coordinates (φ_n, ϑ_n) systematically, it is possible to define a series of different planes passing through the point to be evaluated. For each plane it is possible to calculate the related stress parameters, i.e., $\tau_{n,a}$, $\sigma_{n,min}$, $\sigma_{n,max}$, and $\sigma_{n,a}$. Based on these stress parameters, it is possible to individuate the critical plane having specific spherical coordinates (φ_c, ϑ_c). According to the Matake, the Susmel et al., the Papadopoulos, and the McDiarmid criteria, the critical plane is defined as the plane that displays the maximum value of $\tau_{n,a}$ (Equation (8)).

$$(\varphi_C, \vartheta_C) \rightarrow \max_{\varphi,\vartheta}\{\tau_{n,a}(\varphi, \vartheta)\} \tag{8}$$

Conversely, according to the Findley criterion, the critical plane is defined as the plane that presents the maximum value of the damage parameter (DP) defined as in Equation (9). This is a function of the alternating tangential stress $(\tau_{n,a})$ and the maximum stress reached in a cycle $(\sigma_{n,max})$. Therefore, the Findley criterion could lead to a critical plane having a different orientation with respect to the critical plane according to the other fatigue criteria.

$$(\varphi_C, \vartheta_C) \rightarrow \max_{\varphi,\vartheta}\left\{\tau_{n,a}(\varphi, \vartheta) + \frac{2r_{\tau/\sigma} - 1}{2\left(\sqrt{r_{\tau/\sigma} - r_{\tau/\sigma}^2}\right)}\sigma_{n,max}(\varphi, \vartheta)\right\} \tag{9}$$

where $r_{\tau/\sigma}$ is the ration between the material fatigue limit at symmetrical alternating torsional loading (τ_f) and material fatigue limit at symmetrical alternating bending loading (σ_f) as in Equation (10). It is worth noticing that these material properties can be estimated through simple fatigue tests.

$$r_{\tau/\sigma} = \tau_f/\sigma_f \tag{10}$$

Once the critical plane (φ_c, ϑ_c) has been identified, the various criteria require that the damage parameter on this plane be calculated. In the present paper, the stress parameters related with the critical plane are labeled with the subscript c, i.e., $\tau_{c,a}$, $\sigma_{c,max}$, $\sigma_{c,a}$.

The various criteria differ on how the damage parameter (DP) is calculated. According to Findley, the damage parameter $(DP_{Findley})$ is defined as in Equation (11). According to the Matake criteria, the DP_{Matake} is affected by the alternating (tangential) stress $\sigma_{c,a}$ $(\tau_{c,a})$ (Equation (12)). The Susmel et al. criteria requires calculating the damage parameter $(DP_{Susmel\ et\ al.})$ according to Equation (13). With respect to the criteria proposed by McDiarmid, it is necessary to consider the ultimate tensile stress σ_R (Equation (14)).

$$DP_{Findley} = \tau_{c,a} + \frac{2r_{\tau/\sigma} - 1}{2\left(\sqrt{r_{\tau/\sigma} - r_{\tau/\sigma}^2}\right)}\sigma_{c,max} \tag{11}$$

$$DP_{Matake} = \tau_{c,a} + (2r_{\tau/\sigma} - 1)\sigma_{c,a} \tag{12}$$

$$DP_{Susmel\ et\ al.} = \tau_{c,a} + \left(\tau_f - \frac{\sigma_f}{2}\right)\frac{\sigma_{c,max}}{\tau_{c,a}} \tag{13}$$

$$DP_{McDiarmid} = \tau_{c,a} + \frac{\tau_f}{2\sigma_R}\sigma_{c,max} \tag{14}$$

To implement the Papadopoulos' criteria, it is important to define the maximum octahedral stress $\sigma_{h,max}$ (in the time window T). It can be calculated through Equation (15), where σ_O is a vector with the principal stresses, i.e., the stresses that, for the same time instant t, satisfies Equation (16). $\bar{\bar{I}}$ is the identity matrix. The $DP_{Papadopoulos}$ can be calculated according to Equation (17).

$$\sigma_{h,max} = \max_{T}\left\{\frac{1}{3}\sum_{i=1,2,3}\sigma_{O_i}\right\} \tag{15}$$

$$\det\left|\bar{\bar{\sigma}}(t) - \sigma_O\bar{\bar{I}}\right| = 0 \tag{16}$$

$$DP_{Papadopoulos} = \tau_{c,a} + \left(\frac{3}{2}(2r_{\tau/\sigma} - 1)\right)\sigma_{h,max} \tag{17}$$

Eventually, each fatigue criteria (based on critical plane) state that the component works safely as long as the value of the damage parameter, in each point, is below a specific threshold. Therefore, it is possible to calculate a safety factor (S_F) (for each criterion and in each position) which formulation depends on the implemented criterion. For example, $S_{F_{Findlay}}$ is defined in Equation (18). In Equations (19)–(22) the S_F for the Matake, Susmel et al., McDiarmid, and Papadopoulos criteria can be found, respectively. $S_F > 1$ means that the analyzed stress state has not reached the critical value according to the studied criterion (and vice versa for $S_F < 1$).

$$S_{F_{Findlay}} = \frac{\dfrac{\tau_f}{2\left(\sqrt{r_{\tau/\sigma} - r_{\tau/\sigma}^2}\right)}}{DP_{Findley}} \tag{18}$$

$$S_{F_{Matake}} = \frac{\tau_f}{DP_{Matake}} \tag{19}$$

$$S_{F_{Susmel\ et\ al.}} = \frac{\tau_f}{DP_{Susmel\ et\ al.}} \tag{20}$$

$$S_{F_{McDiarmid}} = \frac{\tau_f}{DP_{McDiarmid}} \tag{21}$$

$$S_{F_{Papadopoulos}} = \frac{\tau_f}{DP_{Papadopoulos}} \tag{22}$$

3. Materials and Methods

In the present paper, two different gear-samples geometries subjected to STBF loading have been modeled by means of Finite Element Model (FEM). In Table 1, the geometrical parameters of the gears are reported. The above-mentioned gears were studied experimentally in [38,40]. The authors have collected experimental images of cracks in several teeth (related to the study conducted in [38,40]) in which failure by TBF occurred. Through these images, it has been possible to extrapolate the position of the nucleation point of the cracks in the teeth root region and the direction (angle) of early propagation (Section 3.1). FEM results have been elaborated through critical plane criteria to characterize the crack behavior in the ρ_{fP} (Section 3.2). The numerical and experimental results have been compared in Section 4.

Table 1. Geometrical parameters of the simulated gears—Gear A [38], Gear B [40].

Description	Symbol	Unit	Gear A	Gear B
Normal module	m_n	mm	3.77301	2.2
Normal pressure angle	α_n	°	22.5	17
Number of teeth	z	-	32	30
Face width	b	mm	15	20
Profile shift coefficient	x	-	0.0681	0.25
Dedendum coefficient	h_{fP}^*	-	1.3153	1.675
Root radius factor	ρ_{fP}^*	-	0.36	0.368
Addendum coefficient	h_{aP}^*	-	1.1595	1.361

3.1. Individuation of Cracks Characteristic through Experimental Images

The crack propagation just after nucleation can be characterized by the two parameters χ and β (Figure 4). Using these coordinates, it has been assumed that the early propagation plane is always perpendicular to the view in Figure 4.

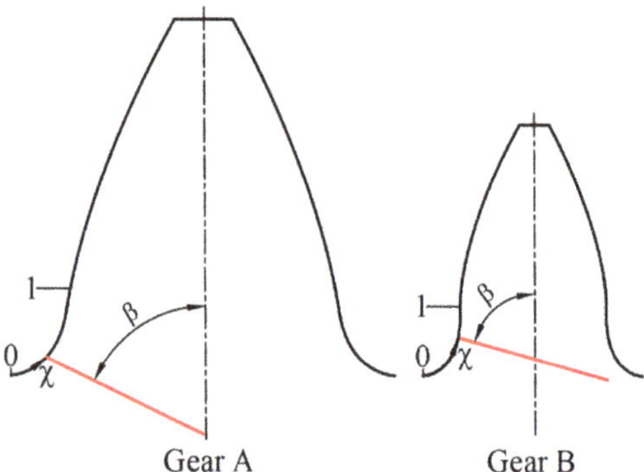

Figure 4. Cracks characteristics (χ, β) in experimental tests.

- χ is a linear coordinate along the ρ_{fP}. This coordinate can take any value from 0 (i.e., lower point in the radius at the foot) to 1 (i.e., connection point between the ρ_{fP} and tooth flank). Through χ, it is possible to define the position of each nucleation point.
- β is the angle between the tooth axis and crack direction in its early propagation.

In Figure 4, χ and β are reported for generic cracks (highlighted in red) in Gear A and Gear B. In the figure, the point in which χ assumes its minimum and its maximum value are indicated.

In Figures 5 and 6, experimental images on which χ and β have been identified are shown. More specifically, Figure 5 shows six images of different teeth belonging to Gear A. The same, referred to as Gear B, is shown in Figure 6.

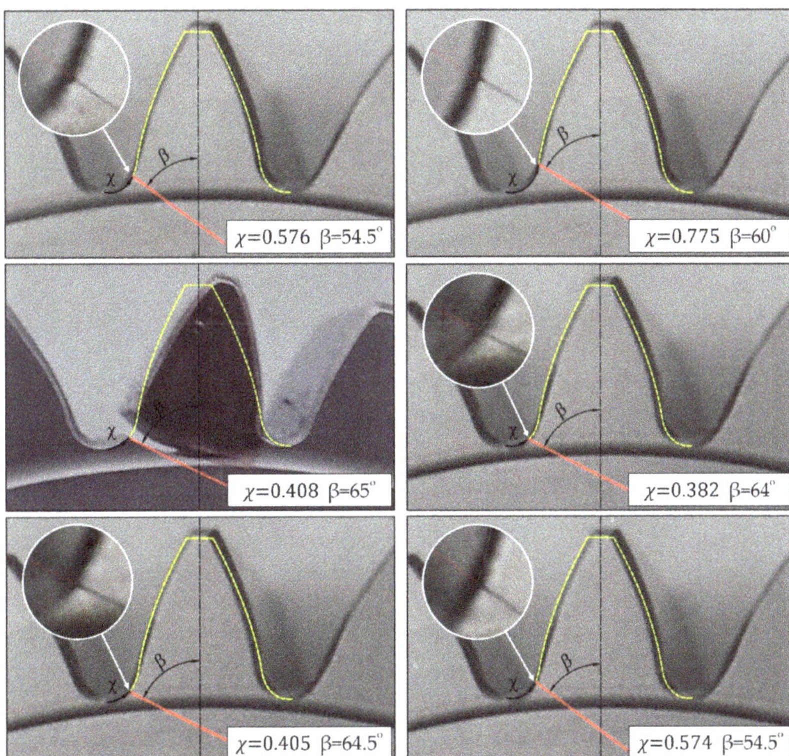

Figure 5. Individuation of cracks characteristics (χ, β) in experimental tests performed on Gear A.

In the figures, the yellow dashed line represents the tooth profile (before the test) while the red solid line represents the direction of early propagation of the crack. It is worth noting that while in Gear B the crack always led to the complete detachment of the tooth, as far as Gear A is concerned, the tests were interrupted when the crack was detected via the variation of the stiffness of the system (even if it did not lead to the complete breakage of the tooth). Therefore, in some images of Gear A, the crack is of limited size and is hidden by the red line, which, however, represents its initial propagation direction.

With respect to Gear A (Figure 5), all the cracks nucleated in $0.382 \leq \chi \leq 0.775$ having a direction $54.5° \leq \beta \leq 65°$. With respect to Gear B (Figure 6), all the cracks nucleated in $0.550 \leq \chi \leq 0.664$ having a direction $42° \leq \beta \leq 51.5°$. It is interesting to notice that, for Gear A, three cracks nucleated in the proximity of $\chi = 0.400$, while the other three cracks nucleated in different points. The latter cracks may be nucleated at different locations due to micro defects in the material. Moreover, in Gear B, the nucleation points have a lower dispersion, but are located in the proximity of the end of the grinding zone where, most likely, a micro notch has formed between the root radius and the beginning of the involute tooth profile.

3.2. Numerical Elaboration Aimed to Characterize Cracks within Tooth Root Radius

The FEM has been set up into the open-source software, Salome-Meca/Code_Aster. In Figures 7 and 8, it is possible to see the STBF test modeling for Gear A and Gear B, respectively. In the present study, 3D simulations have been performed to also consider the boundary effects.

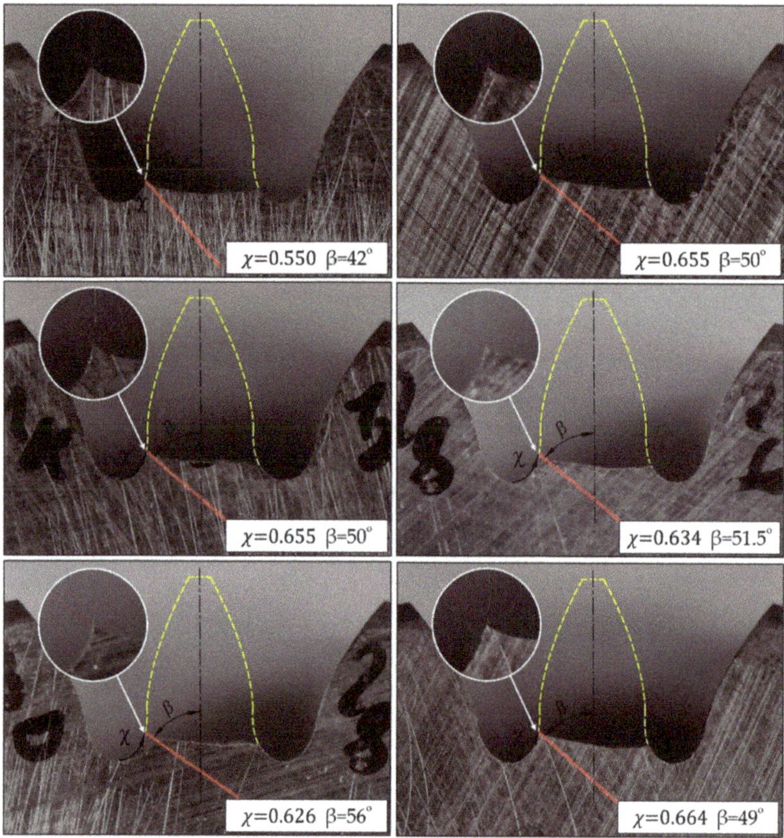

Figure 6. Individuation of cracks characteristics (χ, β) in experimental tests performed on Gear B.

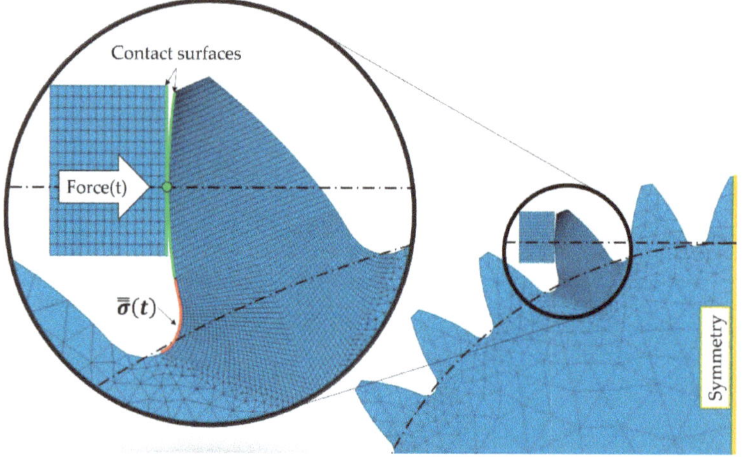

Figure 7. Finite Element Model of the STBF of Gear A.

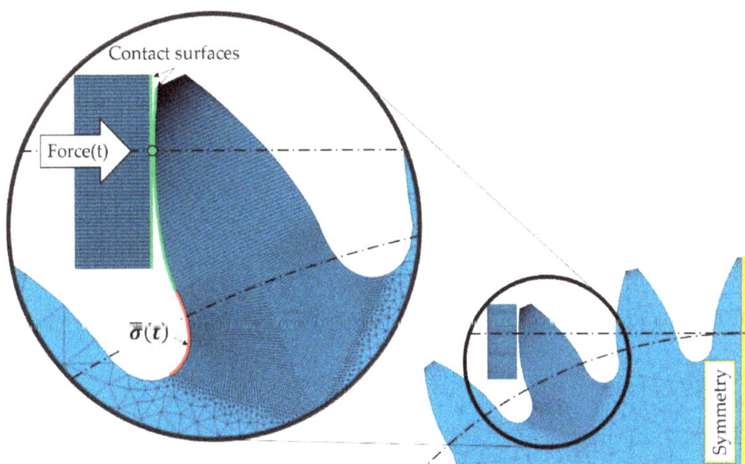

Figure 8. Finite Element Model of the STBF of Gear B.

To reduce the computational effort, only a quarter of each gear has been modeled exploiting symmetries. More specifically, on the one hand, half of the face width has been modeled. On the other hand, gears are symmetric on a plane parallel to the contact-face of the anvil and positioned at half of the Wildhaber distance (yellow line in Figures 7 and 8).

The models have been created through extruded meshes. Linear elements having typical isotropic steel properties have been used i.e., a Young modulus equal to 205,000 MPa and a Poisson's ratio of 0.3. In each model, hexahedral elements have been exploited to model the loaded tooth while TRIA6 elements, i.e., triangular base prisms, have been used to model the remaining volume of the gear. The mesh density has been increased in the loaded tooth after a sensitivity analysis. More specifically, the mesh density has been increased by 10% until the results of the simulations present a variation of less than 1%. The final models have the mesh characteristics listed in Table 2.

Table 2. Mesh characteristics of the simulated gears.

Characteristics	Unit	Gear A	Gear B
Total nodes	#	75,824	214,160
Hexahedral elements	#	54,000	156,690
TRIA6 elements	#	27,855	83,205
Element in the half-face width	#	15	15
Nodes in the tooth flank	#	480	800
Nodes in the tooth root radius	#	496	816

Non-linear simulations have been performed to simulate the contact between the anvil and the tooth flank for each gear. While the analyses are non-linear due to the contacts, the state of stress never exceeded the yielding. In Figures 7 and 8, the contact faces are indicated with green lines and the theoretical contact point is indicated with a green circle. It is located in the intersection between the horizontal line tangent to the base circle (represented in the figures) and the tooth flank. With respect to Gear A, a pulsating compressive force varying sinusoidally from a minimum value of 3700 kN to a maximum value of 37,000 kN has been applied to the anvil. With respect to Gear B, the minimum and maximum value of the force applied result 1498 kN and 14,980 kN, respectively. Through the above-mentioned loading configuration, taking into consideration the symmetries exploited, it has been possible to replicate the experimental conditions, i.e., ratio between the minimum and maximum force of 0.1 (applied in the experimentation). Those levels of force are the loads that averagely lead to a failure in 10^6 cycles.

The stress tensor $\overline{\overline{\sigma}}(t)$ was extracted for both gears in the most critical areas where fracture is expected to nucleate, i.e., within the ρ_{fP} (nodes highlighted with the red line in Figures 7 and 8). At this point, the approaches presented in Section 2 have been applied by defining the material properties $(\sigma_f, \tau_f, \sigma_R)$. In particular, Gear A has been manufactured with VAR 9310 having a bending fatigue limit $\sigma_f = 1400$ MPa, a torsional fatigue limit $\tau_f = 1100$ MPa, and an ultimate tensile strength $\sigma_R = 2700$ MPa. On the other hand, Gear B has been manufactured through 20MnCr5 having $\sigma_f = 516$ MPa, $\tau_f = 303$ MPa, and $\sigma_R = 1028$ MPa.

Therefore, for each gear and for each point within the ρ_{fP}, it has been possible to elaborate the stress tensor $\overline{\overline{\sigma}}(t)$ implementing the different fatigue criteria presented in the previous section. In the present paper, the studied points are the nodes of the computational grid belonging to the ρ_{fP}. This choice was made in order to avoid the need of interpolation. The workflow followed is graphically explained in Figure 9. For each gear, the workflow is structured with four FOR loops. The innermost one analyses data for each simulated time step (in these cases $T = 40$). The FOR loops on ϑ and φ aim to discretize the space by defining the direction of different planes varying by 0.5° each cycle (from 0° to 180°). The FOR loop on the nodes within the ρ_{fP} i.e., $N_{max} = 31$ for Gear A and $N_{max} = 51$ for Gear B, aims to study the most critical positions. Indeed, for each node $N(\theta_c \varphi_c)$, belonging to the symmetry section of the tooth (i.e., the most critical), the critical plane has been individuated through the presented framework. This allowed for achieving a twofold objective. First, it allows us to calculate the damage parameters for each node and each criteria (through Equations (11)–(14) and (17)). Therefore, it has been possible to calculate S_F for each node and each criteria (Equations (18)–(22)) (green boxes in Figure 9). In this way, it has been possible to estimate the differences between nodes in terms of criticality. Moreover, the most critical node according to the different criteria implemented has been established. Second, it has allowed us to identify the direction of the crack propagation (at least in in the proximity of the studied nodes) if it nucleates in any of them (by differentiating between the various fatigue criteria) (blue boxes in Figure 9).

The above-mentioned direction of the critical plane corresponds to the direction of early propagation of the crack after nucleation (evaluated for each node and each criterion). In addition, S_F is representative of the criticality of the node (according to the criteria in question). The combination of these two results, i.e., direction of critical plane and S_F, allowed for obtaining an overview of possible crack propagation scenarios in the ρ_{fP} according to the various criteria. These results have been compared with the experimental ones in terms of crack positions and paths observed after performing STBF tests. The comparison has allowed for assessing the effectiveness of each criterion to correctly predict the failure.

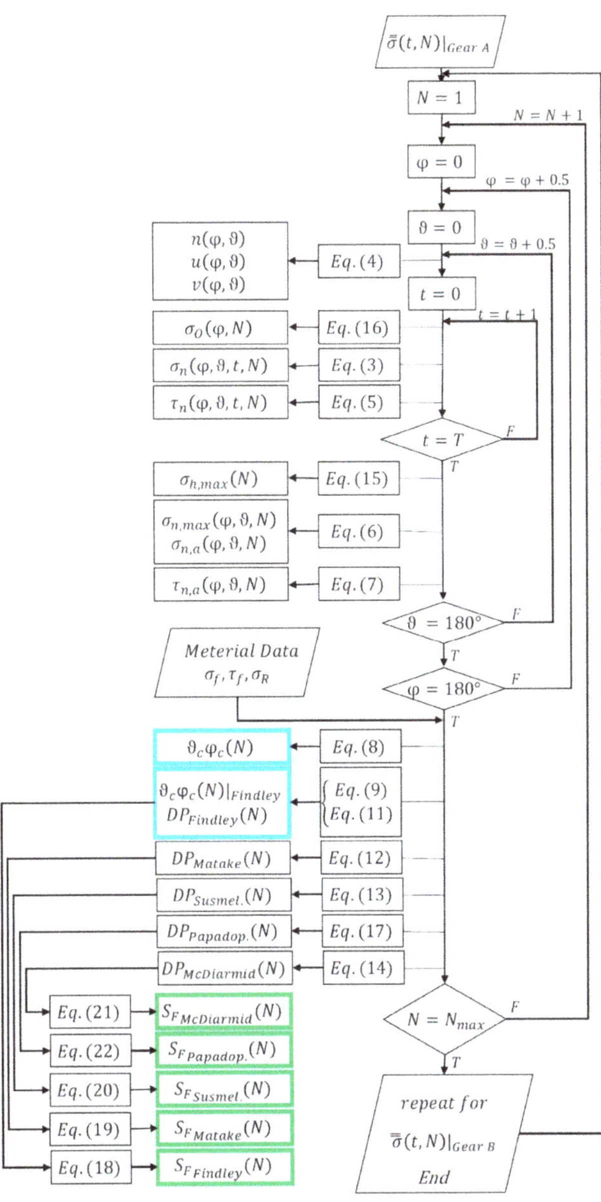

Figure 9. Framework to elaborate the time-dependent stress tensor on each node implementing different fatigue criteria.

4. Results

As mentioned in the previous section, numerical and experimental results have been compared. On the one hand, in STBF specimens, it has been possible to identify both the point where the crack nucleated and the direction of crack propagation for each tooth that failed during the test. On the other hand, through the elaboration of numerical results, for each node within the ρ_{fP} it has been possible to evaluate the damage parameter (it indicates the criticality of the node in question) and the direction of the critical plane (it indicates the direction of the initial crack propagation if the crack nucleates in the studied node).

The comparison has allowed for assessing the effectiveness of each criterion to correctly predict the failure. More specifically, each criterion has been evaluated based on its attitude to:

1. Provide a S_F consistent with the experimental measurements, i.e., how the S_F is close to 1 since the simulated loading condition, according to the experimental results, should lead to a maximum tensile stress σ_F equal to the permissible bending one σ_{FP};
2. Identify the actual critical node, i.e., how close the numerically identified critical node is to the crack nucleation point obtained through experimental tests;
3. Determine the actual crack direction, i.e., how the numerically calculated critical plane direction (at the node closest to nucleation point) is similar to the experimentally observed crack propagation direction.

In Table 3, the minimum S_F calculated according with the investigated criteria and the relative node location is reported for the two gears. With respect to the parameter χ, all the criteria show a congruence in identifying the critical node ($\chi = 0.400$ for Gear A and $\chi = 0.435$ for Gear B). In addition, it is interesting to highlight that, according to the standard [14], the critical node should be located in $\chi = 0.508$ for Gear A and $\chi = 0.430$ for Gear B (the standard [14] defines the critical point as the point of the fillet tangent to a straight line having 30° inclination with respect to the axis of the tooth). Therefore, on the one hand, numerical results lead to individuate the critical node in the same position of the standard for Gear B and in a different position for Gear A. On the other hand, experimental results show a greater variability in the nucleation point that, in turn, are not in agreement with the standard in either case but in very good agreement with the numerical results for Gear A.

Table 3. Minimum S_F calculated through different fatigue criteria and associated critical node location χ.

Gear	$S_{F\,Findley}(\chi)$	$S_{F\,Matake}(\chi)$	$S_{F\,Susmel}(\chi)$	$S_{F\,Papadopoulos}(\chi)$	$S_{F\,McDiarmid}(\chi)$
A	1.08(0.400)	1.96(0.400)	0.79(0.400)	1.13(0.400)	2.14(0.400)
B	0.98(0.435)	1.23(0.435)	0.94(0.435)	0.95(0.435)	1.08(0.435)

With respect to the value of S_F, in Table 3 it emerges that the implementation of the Findley criterion leads to values of S_F closer to the unity for both the gears. Comparable values emerge even when implementing the Papadopoulos criterion. Moreover, while for Gear B all S_F values are close to unity (ranges from 0.94 to 1.23), for Gear A Matake and McDiarmid criteria lead to very high values of S_F i.e., 1.96 and 2.14, respectively. In both the cases it is possible to assert that the Susmel criteria is the most conservative one.

In Figures 10 and 11, experimental and numerical results are graphically compared. Figure 10 is related to the ρ_{fP} of Gear A, Figure 11 concerns the ρ_{fP} of Gear B. In particular, for each of the criteria investigated, the direction of the critical plane calculated in different nodes of the fillet are shown through blue lines. The length of the segments is proportional to the damage parameter. The thicker blue line represents the critical plane having the higher damage parameter. The red lines represent the experimental results and have length as if it was a critical plane having a unit S_F. For each criterion, only the ρ_{fP} and the tooth axis have been reported. The results can be represented graphically in 2D since the critical planes are all perpendicular to the views in the figures.

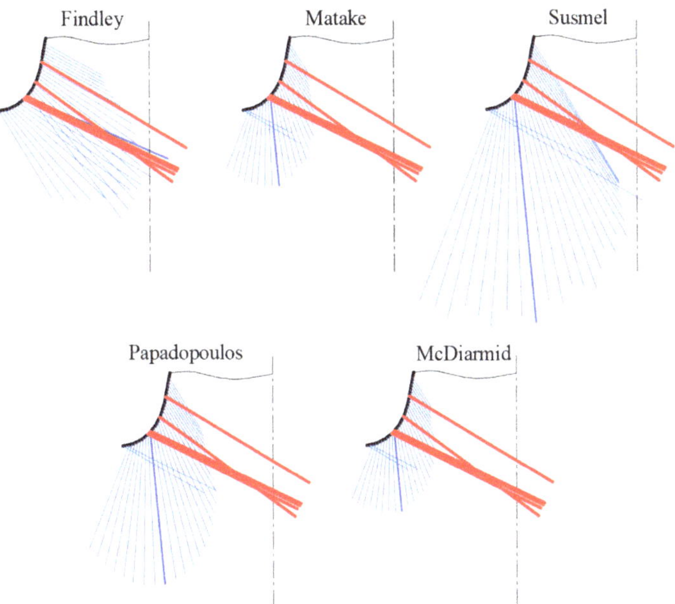

Figure 10. Direction of the critical planes according to the different criteria studied at different nodes for Gear A. Numerical results in blue (segment length proportional to the damage parameter) and experimental results in red (segment length proportional to the damage parameter that lead to a unitary S_F).

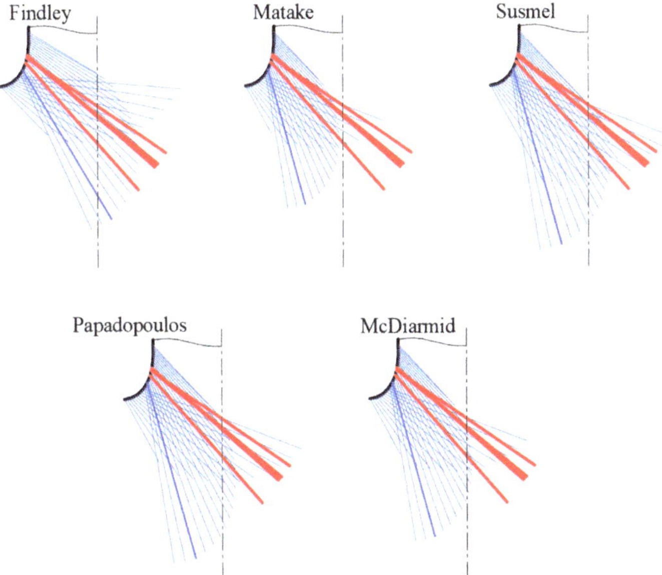

Figure 11. Direction of the critical planes according to the different criteria studied at different nodes for Gear B. Numerical results in blue (segment length proportional to the damage parameter) and experimental results in red (segment length proportional to the damage parameter that lead to a unitary S_F).

Naturally, the direction of the critical planes only varies between Findley and the other criteria, which, in turn, identify the critical plane in the same way. What changes between the various criteria is the value of the damage parameter associated with each node and, therefore, the length of the blue segments.

With respect to Gear A, it is possible to notice that most of the experimentally measured cracks are located in the proximity of the most critical node (i.e., the intersection between the thickest blue line and the radius). However, only the Findley criterion is capable of identifying, with very good approximation, the direction of early propagation of the crack. In addition, Findley's criterion also allows for identifying the direction of cracks even when these nucleate in different positions of the radius, i.e., $\chi = 0.576$, $\chi = 0.574$, $\chi = 0.775$ (most likely due to minor manufacturing or material defects in those positions).

The other criteria lead to very different angles with respect to the ones observed experimentally, e.g., $59°$ difference between the plane with the maximum $\tau_{c,a}$ and the crack observed in its proximity. Therefore, the Findley criterion is the criterion that better models crack nucleation (and early propagation) at the ρ_{fP} of Gear A.

With respect to Gear B, most of the experimentally observed cracks are not in the proximity of the most critical plane calculated numerically. However, also in this case, the Findley criterion is capable of better estimating the crack propagation direction within the whole ρ_{fP}. Indeed, the other criteria suffer from errors ranging from $15°$ to $25°$ while Findley approximates the direction with an error of less than $5°$.

5. Discussion and Conclusions

In the present paper, a methodological approach for implementing five different fatigue criteria based on the critical plane is presented. This relies on the elaboration of the stress tensor $\overline{\overline{\sigma}}(t)$ calculated via FE simulations on specific nodes modeling the ρ_F in STBF loading condition. With the aim of evaluating the accuracy of the different criteria, two different gear geometries have been studied. In both cases, the gears had been experimentally tested and, therefore, it has been possible to obtain the force values leading to the permissible stress, the crack nucleation points, and the crack propagation (just after nucleation) directions in multiple tests. Therefore, the numerical results have been compared with the experimental ones in terms of: (1) capability of the criteria to provide a S_F equal to one; (2) identify the actual critical node; and (3) determine the actual crack direction.

With respect to the point (1), Findley and Papadopoulos are the criteria that lead to the expected outcome most effective in both gears. The Matake criterion leads to overestimating the material strength in both the gears. The Susmel et al. criterion tends to underestimate the material properties and, therefore, it results in being a conservative criterion. The McDiarmid criterion leads to two different results in the two gears, i.e., in Gear B the value of S_F is close to unity while it is more than double for Gear A. This may be due to the high tensile strength of the Gear A material that, in turn, it is considered in the formulation of the damage parameter according to McDiarmid. Eventually, in terms of the point (1), the Findley and Papadopoulos criteria are the most appropriate ones to be applied on gears for estimating fatigue behavior.

With respect to the point (2), all criteria agree in identifying the most critical node in both gear geometries. However, the comparison with experimental results shows that in Gear A, numerical results correctly identify the nucleation point of the crack, while in Gear B, the nucleation point is not accurately identified. Nevertheless, the numerical results of Gear B agree with the standard [14]. In addition, it is worth noting that some cracks in Gear A are located in different points, probably due to micro defects of the material or in the manufacturing process, while in Gear B all the cracks nucleate in the proximity of the end of the grinding zone between the ρ_{fP} and the tooth flank. Moreover, it is possible to notice that the difference in the damage parameter between neighboring nodes is relatively low (less than 3% of difference in the proximity of the most critical node). Therefore, it is possible to state that about 25% of the studied area is subject to a damage parameter above 90% of the maximum damage parameter. Eventually, it is possible to assert that the experimental cracks occur in different positions due to phenomena related to micro-defects that, in turn, were not reproduced with the present FE modeling.

With respect to the point (3), Findley's criterion is undoubtedly the most appropriate for identifying the direction of crack propagation in each possible nucleation point of both the studied gears. Therefore, it is possible to assert that the crack propagation direction at the ρ_{fP} does not follow the plane of maximum alternating shear stress but the plane of maximum damage parameter according to Findley. Indeed, all experimentally identified cracks follow a direction relevant to that indicated by the implementation of Findley's criterion. This result could open the door to the development of new fatigue criteria based on the critical plane for the study of gear. Indeed, an interesting future research direction would be to formulate and/or verify criteria defining the critical plane by the damage parameter (as currently done by Findley's criterion) and, therefore, taking into account also the stress normal to the critical plane for its definition.

Eventually, it is worth noting that the method proposed in this paper has a general validity since it models three-dimensional geometries. However, in the specific case studied in this article, two-dimensional models could also be used to speed up the simulations. In this case, to implement the elaboration of the stress history, it would have been possible to use cylindrical coordinates by setting an angle constant consistent with the simulated model.

Author Contributions: Conceptualization, F.C. and L.M.; methodology, F.C. and L.M.; software, L.F.; validation, L.B., L.M. and F.C.; formal analysis, F.C. and L.M.; investigation, F.C. and L.M.; resources, L.B., F.C. and L.M.; data curation, L.F., F.C. and L.M.; writing—original draft preparation, L.M.; writing—review and editing, F.C.; visualization, L.M., L.B.; supervision, F.C. All authors have read and agreed to the published version of the manuscript.

Funding: This work was supported by the Open Access Publishing Fund of the Free University of Bozen-Bolzano.

Institutional Review Board Statement: Not applicable.

Informed Consent Statement: Not applicable.

Conflicts of Interest: The authors declare no conflict of interest.

Abbreviations

Nomenclature

TBF	Tooth Bending Fatigue
TBS	Tooth Bending Strength
STBF	Single Tooth Bending Fatigue
FE	Finite Element
MCC	Minimum Circumscribed Circle
DP	Damage Parameter
ρ_{fP}	Tooth Root Radius
σ_F	Maximum tensile stress
σ_{FP}	Permissible bending stress
σ_{Flim}	Material strength
$\overline{\overline{\sigma}}(t)$	Stress tensor history
P_n	Stress exerting on a plane defined by a normal vector n
φ_n, ϑ_n	Spherical coordinates of the plane defined by a normal vector n
σ_n	Stress component normal to the plane defined by a normal vector n
τ_n	Stress component tangential to the plane defined by a normal vector n
$\sigma_{n,min}$	Minimum value assumed by σ_n
$\sigma_{n,max}$	Maximum value assumed by σ_n
Γ_n	Curve determined by τ_n along the time
$\tau_{n,a}$	Alternating tangential stress on the plane defined by a normal vector n
$\tau_{n,m}$	Average tangential stress on the plane defined by a normal vector n

$\sigma_{c,max}$ Maximum stress component normal to the critical plane
$\tau_{c,a}$ Alternating tangential stress on the critical plane
σ_f Material fatigue limit at symmetrical alternating bending loading
τ_f Material fatigue limit at symmetrical alternating torsional loading
$r_{\tau/\sigma}$ Ratio between τ_f and σ_f
S_F Safety Factor
χ Linear coordinate along the fillet in the tooth root radius
β Angle between the tooth axis and crack direction
T Time period in a loading cycle
N_{max} Number of nodes modeling the tooth root radius

References

1. Vullo, V. *Gears*; Springer International Publishing: New York, NY, USA, 2020.
2. Radzevich, S.P.; Dudley, D.W. *Handbook of Practical Gear Design*; CRC Press: Boca Raton, FL, USA, 1994.
3. Yadav, A. Different types Failure in gears—A Review. *Int. J. Sci. Eng. Technol. Res.* **2012**, *5*, 82–92.
4. Fernandes, P.J.L.; McDuling, C. Surface contact fatigue failures in gears. *Eng. Fail. Anal.* **1997**, *4*, 99–107. [CrossRef]
5. Li, S.; Kahraman, A. A scuffing model for spur gear contacts. *Mech. Mach. Theory* **2021**, *156*, 104161. [CrossRef]
6. Wu, S.; Cheng, H.S. Sliding wear calculation in spur gears. *J. Tribol* **1993**, *115*, 493–500. [CrossRef]
7. Blake, J.W.; Cheng, H.S. A surface pitting life model for spur gears: Part I—Life prediction. *J. Tribol.* **1991**, *113*, 712–718. [CrossRef]
8. Liu, H.; Liu, H.; Zhu, C.; Zhou, Y. A review on micropitting studies of steel gears. *Coatings* **2019**, *9*, 42. [CrossRef]
9. Fernandes, P.J.L. Tooth bending fatigue failures in gears. *Eng. Fail. Anal.* **1996**, *3*, 219–225. [CrossRef]
10. Pantazopoulos, G.A. Bending fatigue failure of a helical pinion bevel gear. *J. Fail. Anal. Prev.* **2015**, *15*, 219–226. [CrossRef]
11. Davoli, P.; Conrado, E.; Michaelis, K. Recognizing gear failures. *Mach. Des.* **2007**, *63*, 64–67.
12. Bretl, N.; Schurer, S.; Tobie, T.; Stahl, K.; Höhn, B.R. Investigations on tooth root bending strength of case hardened gears in the range of high cycle fatigue. In Proceedings of the American Gear Manufacturers Association Fall Technical Meeting, Indianapolis, IN, USA, 15–17 September 2013; pp. 103–118.
13. Hong, I.J.; Kahraman, A.; Anderson, N. A rotating gear test methodology for evaluation of high-cycle tooth bending fatigue lives under fully reversed and fully released loading conditions. *Int. J. Fatigue* **2020**, *133*, 105432. [CrossRef]
14. *ISO 6336-1:2006, 2006. Calculation of Load Capacity of Spur and Helical Gears, Part 1: Basic Principles, Introduction and General Influence Factors*; ISO: Geneva, Switzerland.
15. *ISO 6336-3:2006, 2006. Calculation of Load Capacity of Spur and Helical Gears, Part 3: Calculation of Tooth Bending Strength*; ISO: Geneva, Switzerland.
16. *ANSI/AGMA 2001-D04, 2004. Fundamental Rating Factors and Calculation Methods for Involute Spur and Helical Gear Teeth*; American Gear Manufacturers Association: Alexandria, Virginia, 2004.
17. Benedetti, M.; Fontanari, V.; Höhn, B.R.; Oster, P.; Tobie, T. Influence of shot peening on bending tooth fatigue limit of case hardened gears. *Int. J. Fatigue* **2002**, *24*, 1127–1136. [CrossRef]
18. McPherson, D.R.; Rao, S.B. Methodology for translating single-tooth bending fatigue data to be comparable to running gear data. *Gear Technol.* **2008**, 42–51. [CrossRef]
19. Dobler, D.I.A.; Hergesell, I.M.; Stahl, I.K. Increased Tooth Bending Strength and Pitting Load Capacity of Fine-Module Gears. *Gear Technol.* **2016**, *33*, 48–53.
20. Concli, F. Tooth Root Bending Strength of Gears: Dimensional Effect for Small Gears Having a Module below 5 mm. *Appl. Sci.* **2021**, *11*, 2416. [CrossRef]
21. McPherson, D.R.; Rao, S.B. Mechanical Testing of Gears. In *Mechanical Testing and Evaluation*; ASM International: Materials Park, OH, USA, 2000; Volume 8, pp. 861–872.
22. Gorla, C.; Conrado, E.; Rosa, F.; Concli, F. Contact and bending fatigue behaviour of austempered ductile iron gears. *Proc. Inst. Mech. Eng. Part C J. Mech. Eng. Sci.* **2018**, *232*, 998–1008. [CrossRef]
23. Concli, F.; Fraccaroli, L. Bending fatigue strength of small size 2 mm module gears. *Wit Trans. Eng. Sci.* **2021**, *133*, 39–46. [CrossRef]
24. Rao, S.B.; McPherson, D.R. Experimental characterization of bending fatigue strength in gear teeth. *Gear Technol.* **2003**, *20*, 25–32.
25. Fontanari, V.; Molinari, A.; Marini, M.; Pahl, W.; Benedetti, M. Tooth Root Bending Fatigue Strength of High-Density Sintered Small-Module Spur Gears: The Effect of Porosity and Microstructure. *Metals* **2019**, *9*, 599. [CrossRef]
26. Winkler, K.J.; Schurer, S.; Tobie, T.; Stahl, K. Investigations on the tooth root bending strength and the fatigue fracture characteristics of case-carburized and shot-peened gears of different sizes. *Proc. Inst. Mech. Eng. Part C J. Mech. Eng. Sci.* **2019**, *233*, 7338–7349. [CrossRef]
27. Argoud, V.; Morel, F.; Pessard, E.; Bellett, D.; Thibault, S.; Gourdin, S. Fatigue behaviour of gear teeth made of case hardened steel: From competing mechanisms to lifetime variability. *Procedia Struct. Integr.* **2019**, *19*, 719–728. [CrossRef]
28. Bonaiti, L.; Rosa, F.; Rao, P.M.; Concli, F.; Gorla, C. Gear root bending strength: Statistical treatment of Single Tooth Bending Fatigue tests results. *Forsch. Ing.* **2021**. [CrossRef]

29. Dixon, W.J.; Mood, A.M. A method for obtaining and analyzing sensitivity data. *J. Am. Stat. Assoc.* **1948**, *43*, 109–126. [CrossRef]
30. Maršálek, P.; Moravec, V. A methodology for gear fatigue tests and their evaluation. *J. Middle Eur. Constr. Des. Cars* **2011**, *9*, 13–17.
31. Conrado, E.; Gorla, C.; Davoli, P.; Boniardi, M. A comparison of bending fatigue strength of carburized and nitrided gears for industrial applications. *Eng. Fail. Anal.* **2017**, *78*, 41–54. [CrossRef]
32. Gorla, C.; Rosa, F.; Concli, F.; Albertini, H. Bending fatigue strength of innovative gear materials for wind turbines gearboxes: Effect of surface coatings. In Proceedings of the ASME International Mechanical Engineering Congress and Exposition, Houston, TX, USA, 9–15 November 2012; Volume 45233, pp. 3141–3147. [CrossRef]
33. Gorla, C.; Rosa, F.; Conrado, E.; Albertini, H. Bending and contact fatigue strength of innovative steels for large gears. *Proc. Inst. Mech. Eng. Part C J. Mech. Eng. Sci.* **2014**, *228*, 2469–2482. [CrossRef]
34. Concli, F.; Bonaiti, L.; Gerosa, R.; Cortese, L.; Nalli, F.; Rosa, F.; Gorla, C. Bending Fatigue Behavior of 17-4 PH Gears Produced by Additive Manufacturing. *Appl. Sci.* **2021**, *11*, 3019. [CrossRef]
35. Gasparini, G.; Mariani, U.; Gorla, C.; Filippini, M.; Rosa, F. Bending fatigue tests of helicopter case carburized gears: Influence of material, design and manufacturing parameters. *Gear Technol.* **2009**, 68–76.
36. Concli, F. Austempered Ductile Iron (ADI) for gears: Contact and bending fatigue behavior. *Procedia Struct. Integr.* **2018**, *8*, 14–23. [CrossRef]
37. Bonaiti, L.; Concli, F.; Gorla, C.; Rosa, F. Bending fatigue behaviour of 17-4 PH gears produced via selective laser melting. *Procedia Struct. Integr.* **2019**, *24*, 764–774. [CrossRef]
38. Gasparini, G.; Mariani, U.; Gorla, C.; Filippini, M.; Rosa, F. Bending Fatigue Tests of Helicopter Case Carburized Gears: Influence of Material, Design and Manufacturing Parameters. In Proceedings of the American Gear Manufacturers Association 369 (AGMA) Fall Technical Meeting, San Antonio, TX, USA, 12–14 October 2008; pp. 131–142.
39. Gorla, C.; Rosa, F.; Conrado, E.; Concli, F. Bending fatigue strength of case carburized and nitrided gear steels for aeronautical applications. *Int. J. Appl. Eng. Res.* **2017**, *12*, 11306–11322.
40. Bonaiti, L.; Bayoumi, A.B.M.; Concli, F.; Rosa, F.; Gorla, C. Gear root bending strength: A comparison between Single Tooth Bending Fatigue Tests and meshing gears. *J. Mech. Des.* **2021**, *143*, 103402. [CrossRef]
41. Concli, F.; Fraccaroli, L.; Maccioni, L. Gear Root Bending Strength: A New Multiaxial Approach to Translate the Results of Single Tooth Bending Fatigue Tests to Meshing Gears. *Metals* **2021**, *11*, 863. [CrossRef]
42. Concli, F.; Maccioni, L.; Bonaiti, L. Reliable gear design: Translation of the results of single tooth bending fatigue tests through the combination of numerical simulations and fatigue criteria. *WIT Trans. Eng. Sci.* **2021**, *130*, 111–122. [CrossRef]
43. Concli, F.; Maccioni, L. Critical planes criteria applied to gear teeth: Which one is the most appropriate to characterize crack propagation? *WIT Trans. Eng. Sci.* **2021**, *133*, 15–25. [CrossRef]
44. Findley, W.N. A theory for the effect of mean stress on fatigue of metals under combined torsion and axial load or bending. *J. Eng. Ind.* **1959**, *81*, 301–305. [CrossRef]
45. Matake, T. An explanation on fatigue limit under combined stress. *Bull. JSME* **1977**, *20*, 257–263. [CrossRef]
46. McDiarmid, D.L. *Fatigue under Out-of-Phase Biaxial Stresses of Different Frequencies. Multiaxial Fatigue*; ASTM International: West Conshohocken, PA, USA, 1985. [CrossRef]
47. Papadopoulos, I.V. A high-cycle fatigue criterion applied in biaxial and triaxial out-of-phase stress conditions. *Fatigue Fract. Eng. Mater. Struct.* **1995**, *18*, 79–91. [CrossRef]
48. Susmel, L.; Tovo, R.; Lazzarin, P. The mean stress effect on the high-cycle fatigue strength from a multiaxial fatigue point of view. *Int. J. Fatigue* **2005**, *27*, 928–943. [CrossRef]
49. Papadopoulos, I.V. Critical plane approaches in high-cycle fatigue: On the definition of the amplitude and mean value of the shear stress acting on the critical plane. *Fatigue Frac. Eng. Mater. Struct.* **1998**, *21*, 269–285. [CrossRef]

 *metals*

Article

Coupling Finite Element Analysis and the Theory of Critical Distances to Estimate Critical Loads in Al6060-T66 Tubular Beams Containing Notches

Marcos Sánchez, Sergio Cicero *, Borja Arroyo and José Alberto Álvarez

LADICIM (Laboratory of Materials Science and Engineering), University of Cantabria, E.T.S. de Ingenieros de Caminos, Canales y Puertos, Av/Los Castros 44, 39005 Santander, Spain; marcos.sanchez@unican.es (M.S.); borja.arroyo@unican.es (B.A.); alvareja@unican.es (J.A.Á.)
* Correspondence: ciceros@unican.es; Tel.: 0034-942-200917

Received: 29 September 2020; Accepted: 17 October 2020; Published: 20 October 2020

Abstract: This paper validates a methodology for the estimation of critical loads in tubular beams containing notch-type defects. The methodology is particularized for the case of Al6060-T66 tubular cantilever beams containing U-shaped notches. It consists in obtaining the stress field at the notch tip using finite element analysis (FEA) and the subsequent application of the theory of critical distances (TCD) to derive the corresponding critical load (or load-bearing capacity). The results demonstrate that this methodology provides satisfactory predictions of fracture loads.

Keywords: critical load; fracture; tubular cantilever beam; U-notch; theory of critical distances

---

1. Introduction

Tubular sections are widely used in engineering applications such as transport systems (e.g., tubes, pipelines), naval and aeronautical engineering, offshore equipment, or lifting systems (e.g., cranes), among others. These structural solutions have been demonstrated to have great strength against different types of loading conditions such as torsion, compression, or multiaxial bending. In addition, their specific shape has proved to be an adequate solution for structures exposed to wind, water, or wave loads, and they are less prone to corrosion processes, as they do not tend to generate local accumulations of water. With all this, it is of great importance from an engineering perspective to be able to estimate the critical loads of this kind of structures, especially when they contain defects.

In the field of structural integrity, the analysis of defects plays an important role in ensuring the safety of structural components. Structural integrity procedures (e.g., [1–3]) are able to evaluate components containing cracks, combining fracture and plastic collapse analyses. However, in many cases, the structures present defects with finite radii on the tip. These defects are generally named notches, and if they are assessed as crack-like defects using standard methodologies (traditionally based on fracture mechanics [4,5]), the results tend to be over-conservative. This is caused by the fact that notches generate more relaxed stress fields at their tip (when compared to those generated by cracks). Apparently, the material develops a higher fracture resistance (usually referred to as the apparent fracture toughness) than that developed in cracked conditions (fracture toughness). Therefore, it is necessary to provide structural assessment methodologies that are capable of taking the notch effect into account, providing accurate predictions of the resulting critical loads. Different works (e.g., [6–11]) dealing with the structural integrity of tubular sections may be found in literature, although failure processes (e.g., plastic collapse, buckling) and defect types (e.g., cracks, cutouts) are different to those considered in this work.

In this sense, when dealing with notch assessments, there are two main types of criteria: the global criterion (based on the use of a notch stress intensity factor, analogously to ordinary fracture mechanics), and local criteria (based on the study of stress or strain fields around the notch tip). Among the latter, the theory of critical distances (TCD) stands out, and its applicability in fracture assessments has been widely reported in the literature for a variety of materials (such as polymers [12,13], metals [14,15], composites [16], or ceramics [17,18]). Moreover, the TCD has also been validated to analyze phenomena such as fatigue [19] or environmentally assisted cracking [20] and has been applied to different length scales [19,21,22].

The TCD is actually a group of methodologies initially proposed in the mid-twentieth century by Neuber [23] and Peterson [24] to predict the fatigue behavior of structural components containing notches. All these methodologies have in common the use of two additional parameters: a material length parameter called the critical distance (L), which is defined by Equation (1), and a material strength parameter named the inherent strength (σ_0). In fracture analysis, both parameters are directly related with the material fracture resistance (K_{mat}) through Equation (1).

$$L = \frac{1}{\pi}\left(\frac{K_{mat}}{\sigma_0}\right)^2 \tag{1}$$

For brittle materials (e.g., ceramics) or quasi-brittle materials (e.g., many fiber-reinforced composites), the inherent material strength is equal or very close to the corresponding ultimate tensile strength (σ_u). Otherwise, σ_0 tends to be higher than σ_u, with this tendency being more pronounced as long as plasticity is developed in the vicinity of the notch. In such cases, σ_0 has to be determined (calibrated) through experimental tests of specimens containing notches with different radii, or through a combination of experimental tests and finite element (FE) modeling.

Within the different approaches proposed by the TCD, the point method (PM) stands out for its simplicity, and provides similar results to other TCD methodologies, such as the line method, the area method, or the volume method, among others [19]. According to the PM criterion, fracture occurs when the stress equates the inherent strength, σ_0, at a distance equal to $L/2$ from the defect tip. The mathematical expression is given by Equation (2):

$$\sigma\left(\frac{L}{2}\right) = \sigma_0 \tag{2}$$

Thus, the PM allows the fracture behavior of notched components to be analyzed by simply knowing L together with the (linear elastic) stress field at the notch tip. The evolution of FE tools allows the stress distribution at a stress concentrator to be more easily determined, something that has allowed extensive validation of the TCD methodologies [19]. However, this validation has been strongly focused on fracture mechanics notched specimens (e.g., CT and single edge notched bend (SENB) samples). In this context, this paper attempts to validate the application of the TCD (coupled with FE analyses) on a larger scale in real structural components (in this case, tubular cantilever beams containing U-notches).

With all of this, Section 2 presents the material and methods, Section 3 gathers the results obtained experimentally and through the TCD-FE analysis, together with the corresponding discussion, and Section 4 summarizes the main conclusions.

2. Materials and Methods

2.1. Materials

The material employed in the present study is a 6060-T66 aluminum alloy. It belongs to the wrought aluminum–magnesium–silicon alloys (6xxx series), and it has been additionally solution heat-treated and artificially aged (T66). This alloy is especially optimized for extrusion processes and develops good tensile strength. It has remarkable corrosion resistance and good weldability. It is commonly used in engineering applications (e.g., railway, automotive industry, building construction, etc.).

In order to carry out the experimental program, four 1.8 m long tubular beams were employed. Two of them had an outer diameter of 312 mm and 6 mm thickness, with the other two beams having an outer diameter of 260 mm and 5 mm thickness. Table 1 gathers the nominal chemical composition of this aluminum alloy [25].

Table 1. Chemical composition (nominal) of Al6060-T66 [25].

Si	Fe	Cu	Mn	Mg	Cr	Zn	Ti	Al
0.30–0.60	0.10–0.30	≤0.10	≤0.10	0.35–0.60	≤0.05	≤0.15	≤0.10	balance

2.2. Methods

The present study proposes a methodology for the analysis of tubular cantilever beams containing (circumferential) through thickness U-notches by applying the TCD. This requires completing experimental tests and FE simulations.

Regarding the experimental program, three of the tubes mentioned above were conducted to failure through bending tests, but previously both fracture and tensile tests were performed in order to characterize the material. Fracture and tensile specimens were machined from the remnant fourth tube (with an outer diameter of 260 mm and 5 mm thickness).

Three tensile tests were conducted according to the ASTM E8M standard [26]. Figure 1 shows the dimensions of the samples that were machined in the longitudinal direction. The tests were carried out with a loading rate of 5 mm/min.

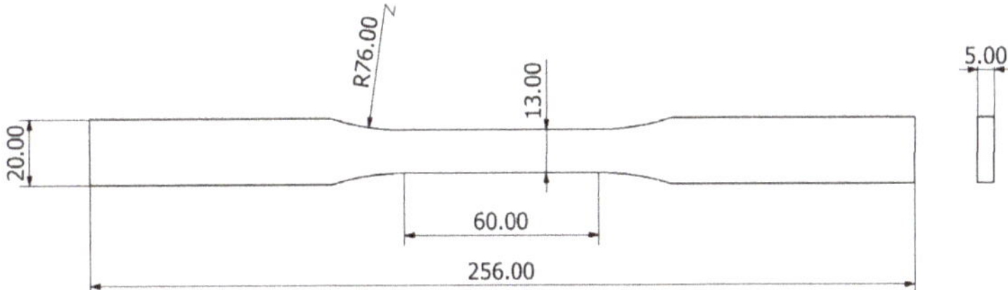

Figure 1. Tensile test specimens. Dimensions in mm.

Subsequently, the fracture behavior of AL6060-T66 was characterized. A total of nine SENB specimens were tested following ASTM E1820 [27]. Three specimens for each notch radii were obtained in LC orientation: the opening stresses act in the longitudinal direction of the pipe, and the defect propagates circumferentially. The notch radii considered in this work are 0 mm (crack-like defect), 1 mm, and 2 mm. Notches of finite radius (1 mm and 2 mm) were obtained by machining, whereas crack-like defects were generated by fatigue pre-cracking according to ASTM E1820. Figure 2 shows a schematic of the specimen used in the fracture characterization. It can be noticed that the width is slightly smaller than the tube thickness due to the need to have prismatic samples. The loading rate was 10 mm/min.

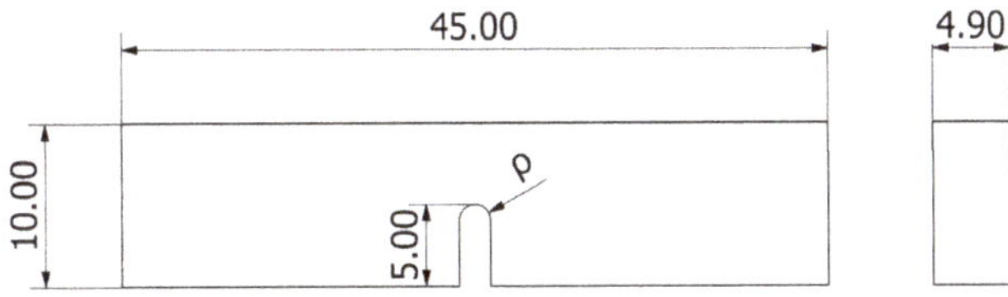

Figure 2. Schematic of fracture single edge notched bend (SENB) specimens. Dimensions in mm.

To conclude with the experimental works, the remnant three tubular cantilever beams were prepared to be tested. Through-thickness circumferential U-notches were machined at a distance of approximately 350 mm from one of the tube ends. In order to obtain a fixed support, the same tube end was introduced 330 mm in reinforced concrete. Figure 3 shows an image of the experimental setup, Figure 4 represents a schematic of the notched tubular cantilever beams, and Table 2 gathers the geometry of both the tubes and the notches.

Figure 3. Experimental setup.

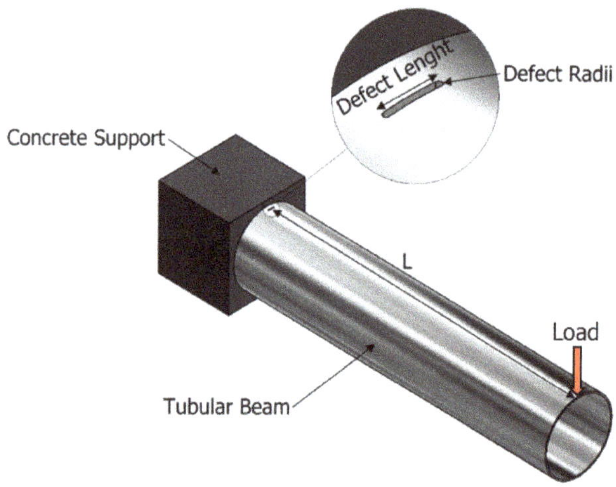

Figure 4. Schematic of the tubular cantilever beams containing a U-notch close to the fixed support.

Table 2. Geometrical parameters of the tubes and their corresponding U-notch: $\varnothing$, outer diameter; B, tube thickness; D, distance from concrete support to notched section; L, distance from applied load to notched section; $2a$, defect length; ρ, defect radius. Dimensions in mm.

Tube	Material	$\varnothing$	B	D	L	$2a$	ρ
AL1	AL6060-T66	312	6.0	30.4	1451	27.2	0.8
AL2	AL6060-T66	312	6.0	27.0	1448	27.2	1.5
AL3	AL6060-T66	260	5.0	21.	1452	45.3	0.8

In order to obtain the experimental critical loads (load-bearing capacity), the tubes were set up in the testing bench, ensuring that the solid concrete block was totally fixed with screws avoiding any kind of movement. A single vertical load was applied at the free edge with a testing rate of 10 mm/min, while a calibrated laser comparator measured the resulting deflection.

As explained above, the application of the PM requires the stress field around the defect tip to be determined. With this aim, FE analyses were carried out. The simulations were performed in linear elastic conditions using the finite element software ANSYS 19.2 (Ansys Inc, Canonsburg, PA, USA) both in the SENB specimens and the cantilever beams.

The simulation of the SENB specimens (Figure 5a) was performed using a structured mesh composed by 20-node hexahedron elements, as shown in Figure 5b. The area surrounding the notch tip was discretized using a much finer mesh, because of the higher stress gradient generated in that zone. For each notch radius, the stress–distance curves were finally obtained in the middle line of the fracture section and for the corresponding average value of the critical loads. The stresses used in the analyses are the corresponding maximum principal stresses, which, in these particular structural conditions, act in the longitudinal direction of the tubular beams.

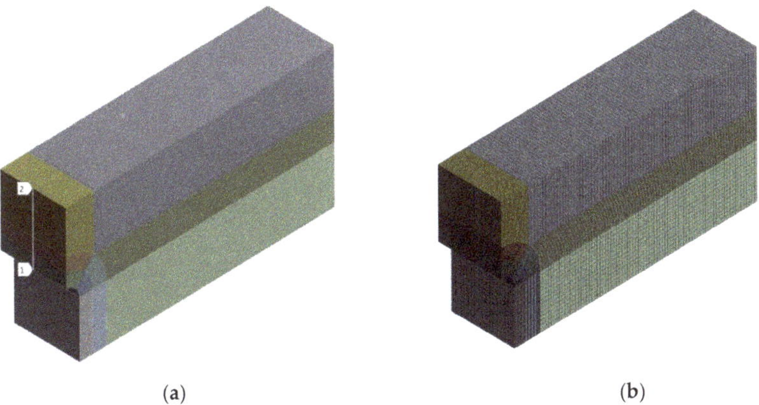

(a)   (b)

Figure 5. Geometry of the model used in finite element (FE) simulations, showing the middle line on the fracture section (**a**) and the generated mesh (**b**).

Once the stress–distance curves for each notch radius were determined, the PM was applied to calibrate the material parameters. When PM is used, it is sufficient to obtain the cutoff point between the different curves, which theoretically corresponds to the coordinates ($L/2$, σ_0), as shown in Figure 6.

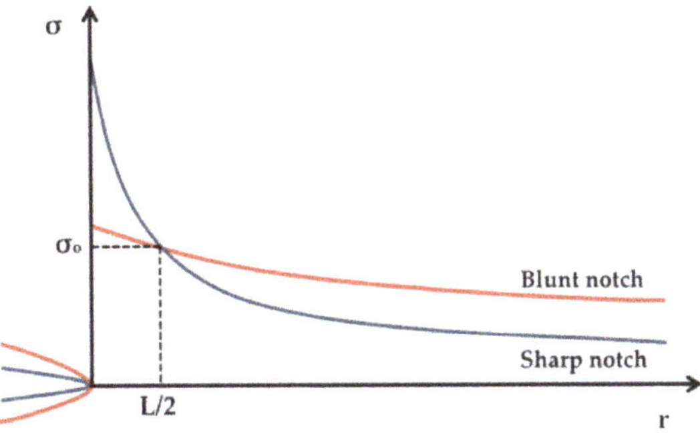

Figure 6. Obtaining theory of critical distances (TCD) parameters using the stress–distance curves.

Finally, the three cantilever tubular beams were modeled (see an example in Figure 7). Again, a structured mesh composed by 20-node hexahedron elements was used. The notch region was partitioned in order to generate a refined mesh, also ensuring 20 elements along the tube thickness. The simulation was performed with just a half of the tube because of the symmetry conditions, applying the load at the free end of each beam. The part of the tube fixed in the concrete block had all the movements restricted, and the points of the tube located in the symmetry plane could only have displacements in such a plane. Here, it is important to notice that the critical load of the complete tubular beam is twice the critical load of the model. A path was created on each tube, starting at the notch root, at half of the tube thickness and along the circumferential direction. Thus, the stress–distance curve (along the corresponding path) was obtained for each tube under the load being applied. Finally, the estimated critical load (or load-bearing capacity) was that for which Equation (2) was satisfied.

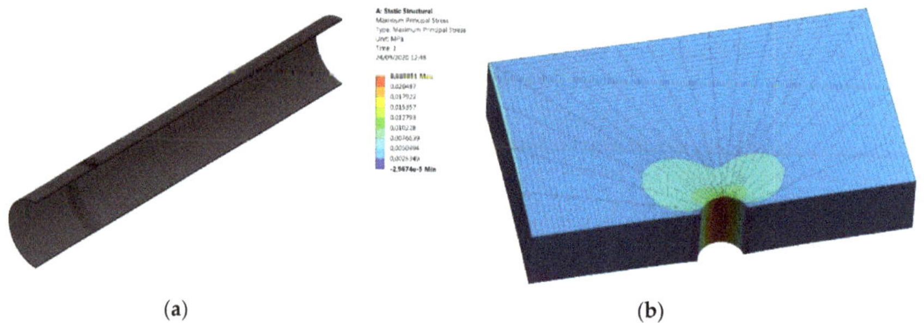

(a)   (b)

Figure 7. (a) Mesh employed in the FEA (finite element analysis) of the tubular beams; (b) detail of the notch tip.

3. Results

The main tensile properties of the Al6060-T66 being used are gathered in Table 3, with E being the Young's modulus, $\sigma_{0.2}$ being the proof strength, σ_u being the ultimate tensile strength, and ε_{max} being the strain under maximum load.

Table 3. Mechanical properties (mean and standard deviation).

Material	E (MPa)	$\sigma_{0.2}$ (MPa)	σ_u (MPa)	ε_{max} (%)
AL 6060	70,750 ± 554	215.0 ± 1.7	264.4 ± 1.8	11.60 ± 0.31

Table 4 presents the experimental results of the fracture tests, with Figure 8 showing some representative examples of the experimental load–displacement curves obtained for each notch radius. Here, it is important to notice that the fracture resistance values obtained are high, even in cracked conditions. In this sense, Equation (3) provides a criterion to estimate the onset of the plane stress conditions [19], with B being the thickness and σ_y being the yield stress (the proof stress for the material being analyzed here). It is straightforward to derive that plane stress conditions are achieved for fracture resistance values above 26.7 MPam$^{1/2}$, approximately, so all the SENB specimens being tested are under plane stress conditions, explaining the high values of fracture resistance obtained here. This is also important to justify the scarce influence of the election of the tube selected for tensile and fracture characterization. As long as the two possible thicknesses (5 mm vs. 6 mm) generate fully plane stress conditions, the influence of this dimension on the resulting fracture resistance may be considered to be negligible.

$$K_{Plane\ Stress} = \sigma_Y(\pi B)^{1/2} \tag{3}$$

Table 4. Experimental results obtained in SENB specimens. K^N_{mat} in cracked specimens correspond to the material fracture toughness K_{mat}.

Material	Specimen	ρ (mm)	Defect Length (mm)	Critical Load (N)	K^N_{mat} (MPa·m$^{1/2}$)
	0-1	0	4.23	1208.8	51.89
	0-2		4.62	1341.6	59.42
	1-1	1	5.00	1235.8	96.53
	1-2		5.00	1236.2	92.62
AL6060-T66	1-3		5.00	1226.7	103.56
	2-1	2	5.00	1296.1	125.49
	2-2		5.00	1259.2	116.47
	2-3		5.00	1259.2	130.03

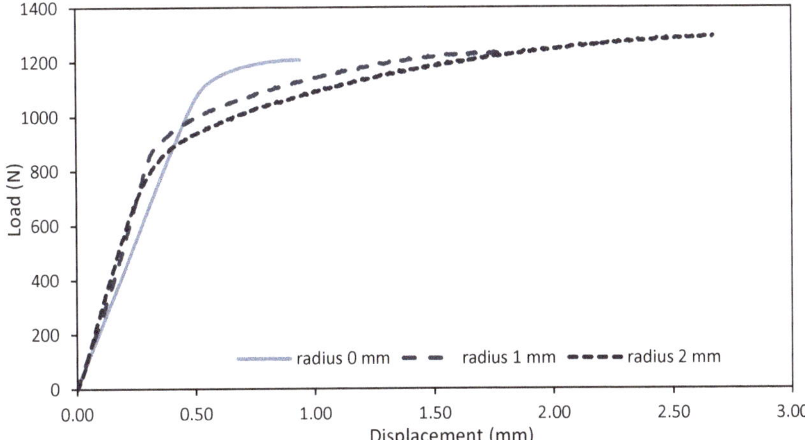

Figure 8. Load–displacement curves of some of the fracture tests.

The load–displacement curves of the structural tests performed on the notched cantilever beams are shown in Figure 9, while the corresponding values of the critical load (in terms of the experimental load-bearing capacity, LBC$_{exp}$) are presented in Table 5.

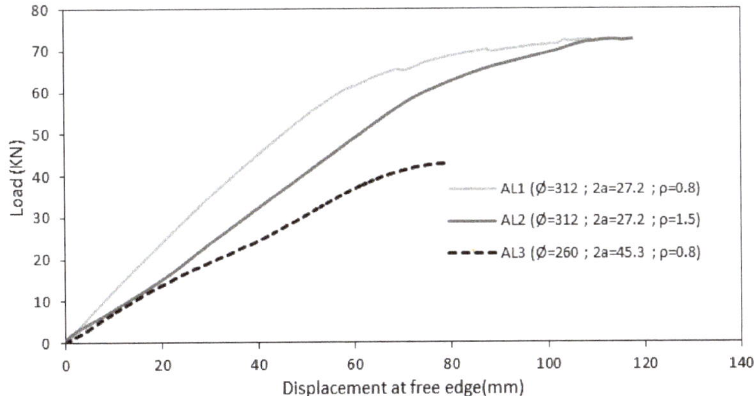

Figure 9. Load–displacement curves of the different tubular beam.

Table 5. Values of L and σ_0 obtained from calibration, together with the experimental and the estimated values of load-bearing capacity (LBC).

Tube	L (mm)	σ_0 (MPa)	LBC$_{exp}$ (kN)	LBC$_{est}$ (kN)
AL1			72.65	76.67
AL2	0.22	920	72.75	87.62
AL3			42.86	39.15

Concerning the FE simulations, the stress–distance curves obtained in the fracture section of the SENB specimens are shown in Figure 10. When the PM is applied in more than two geometries, and the number of tests is limited, it can be observed that the different curves do not necessarily cross each other at the same point, as shown in Figure 6. Following the PM, and considering the inherent scatter of fracture processes, a much larger number of specimens per notch radius would be needed to obtain a single crossing point. For this reason, the material parameters, gathered in Table 5, have been obtained in this work as the average of the different cutoff points, the essential assumption of the PM being reasonably fulfilled. Analogously, Figure 11 shows the stress–distance curves corresponding to the different tubular beams when a load of 1 N is applied in the free edge.

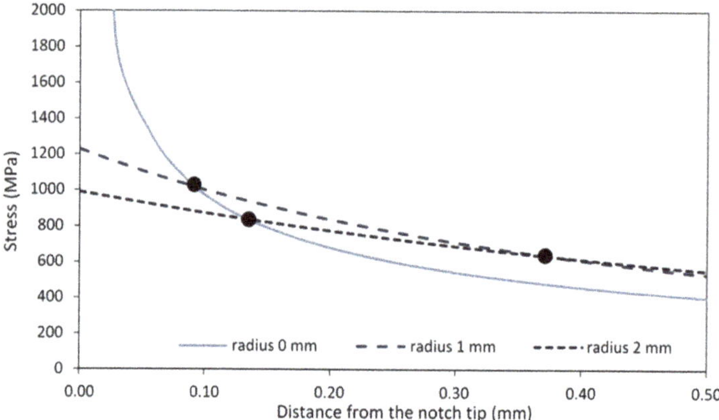

Figure 10. Stress–distance curves at critical load in SENB specimens. The solid circles correspond to the cutoff points.

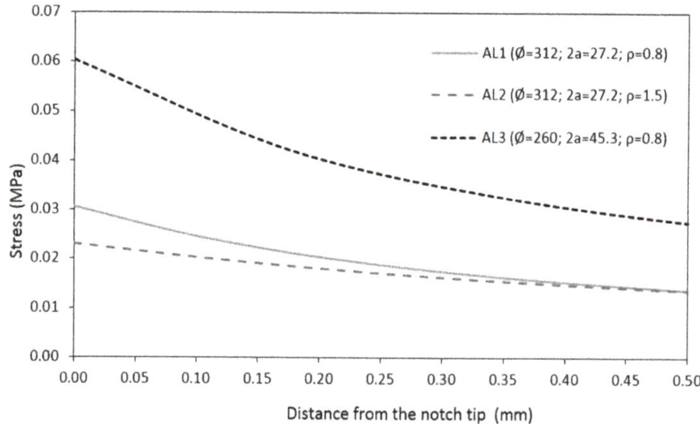

Figure 11. Stress–distance curves in tubular beams when applying a unit load (1 N) at the free end.

Once the experimental results and the FE modeling have been presented, Table 5 also shows the estimations of the load-bearing capacity (LBC$_{est}$) for each tube.

Figure 12 compares the experimental results and the corresponding TCD-FE estimations, showing acceptable predictions of the load-bearing capacity. All the results are basically in the ±20% scatter band, which is generally accepted in fracture research [19,28–30], with an average overestimation of the LBC of +5.7%. This is also understandable, taking into consideration that this approach does not include any safety factor, something commonly used in structural integrity assessments. It can be noticed that the maximum deviation (+20%) occurs in the tube with the largest notch radii ($\rho = 1.5$ mm), which has the more pronounced nonlinear behavior (see Figure 9). It seems that although the TCD compensates nonlinearities with the calibration process (through σ_0 values larger than σ_u), the resulting LBC estimations may lose accuracy when the material's nonlinear behavior becomes more developed.

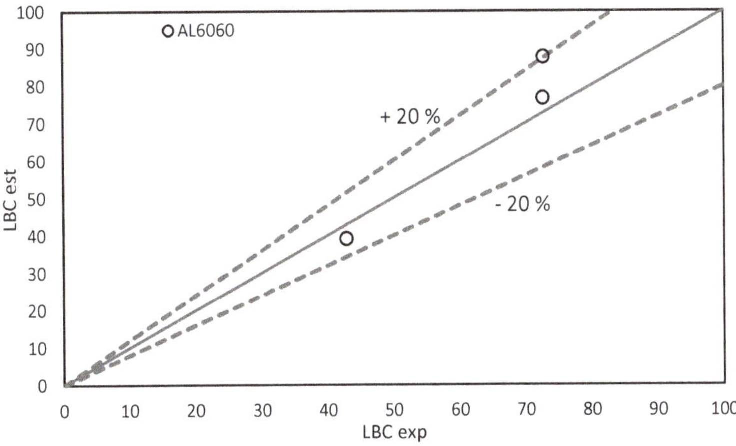

Figure 12. Comparison between the experimental results (LBCexp) and the resulting estimations (LBCest).

4. Conclusions

In this paper, a methodology for the estimation of critical loads in tubular beams containing U-notches has been validated. The methodology is based on the application of the theory of critical distances (TCD) through the point method (PM) and finite element (FE) linear elastic simulations. The methodology has been validated in three Al6060-T66 cantilever beams containing circumferential through thickness U-notches, providing specific validation beyond that reported in fracture mechanics specimens.

Tensile and fracture tests allowed the corresponding material mechanical properties to be estimated, and the combination of fracture tests on notched SENB specimens and FE analyses allowed the material critical distance and inherent strength to be calibrated.

With all this, FE simulations of the notched cantilever beams were performed, determining the estimations of the critical loads (or load-bearing capacities) as those loads for which the PM criterion was fulfilled.

The predicted critical loads represent acceptable estimations of the experimental critical loads, almost within the typical accepted scatter band for fracture processes (± 20%), and with an average overestimation of + 5.7% (without any use of safety factors). The largest deviation (+ 20.4%) was observed in the beam developing the most evident nonlinear behavior, thus being further away from the initial linear elastic nature of the TCD, which makes it harder to compensate by the calibration process of the material critical distance.

The methodology may be applied to other structural components containing any kind of notches. However, further validation is required to establish general conclusions about its accuracy.

Author Contributions: Conceptualization, S.C.; methodology, S.C., M.S., B.A., and J.A.Á.; formal analysis, S.C and M.S.; investigation M.S., S.C., and B.A.; writing—original draft preparation, M.S.; writing—review and editing, M.S., S.C, B.A., and J.A.Á. All authors have read and agreed to the published version of the manuscript.

Funding: This research received funding from the Spanish Ministry of Science and Innovation, grant PGC2018-095400-B-I00 "Comportamiento en fractura de materiales compuestos nano-reforzados con defectos tipo entalla".

Conflicts of Interest: The authors declare no conflict of interest.

References

1. Kocak, M.; Webster, S.; Janosch, J.J.; Ainsworth, R.A.; Koers, R. *FITNET Fitness-for-Service (FFS) Procedure*; GKSS: Hamburg, Germany, 2008; Volume 1.

2. *BS7910, Guide to Methods for Assessing the Acceptability of Flaws in Metallic Structures*; British Standards Institution: London, UK, 2019.

3. *R6: Assessment of the Integrity of Structures Containing Defects, Rev. 4*; EDF Energy: Gloucester, UK, 2015.

4. Anderson, T.L. *Fracture Mechanics: Fundamentals and Applications*, 4th ed.; CRC Press—Taylor and Francis Group: Boca Raton, FL, USA, 2005.

5. Broek, D. *Elementary Engineering Fracture Mechanics*, 4th ed.; Martinus Nijhoff: Dordrecht, The Netherlands, 2012.

6. Stasiewicz, P. Analytical and Experimental Studies of Stability of Cylindrical Shells with a Cut-Out. *Mech. Mech. Eng.* **2013**, *17*, 167–176.

7. Staat, M. Local and global collapse pressure of longitudinally flawed pipes and cylindrical vessels. *Int. J. Press. Vessel. Pip.* **2005**, *82*, 217–225. [CrossRef]

8. Ahn, S.-H.; Nam, K.W.; Takahashi, K.; Ando, K. Comparison of experimental and finite element analytical results for the strength and the deformation of pipes with local wall thinning subjected to bending moment. *Nucl. Eng. Des.* **2006**, *236*, 140–155. [CrossRef]

9. Schenk, C.A.; Schuëller, G.I. Buckling analysis of cylindrical shells with cutouts including random boundary and geometric imperfections. *Comput. Methods Appl. Mech. Eng.* **2007**, *196*, 3424–3434. [CrossRef]

10. Yeh, M.-K.; Lin, M.-C.; Wu, W.-T. Bending buckling of an elastoplastic cylindrical shell with a cutout. *Eng. Struct.* **1999**, *21*, 996–1005. [CrossRef]

11. Alashti, R.A.; Rahimi, G.H.; Poursaeidi, E. Plastic limit load of cylindrical shells with cutouts subject to pure bending moment. *Int. J. Press. Vessel. Pip.* **2008**, *85*, 498–506. [CrossRef]

12. Cicero, S.; Madrazo, V.; Carrascal, I.A. Analysis of notch effect in PMMA using the Theory of Critical Distances. *Eng. Fract. Mech.* **2012**, *86*, 56–72. [CrossRef]

13. Taylor, D.; Merlo, M.; Pegley, R.; Cavatorta, M.P. The effect of stress concentrations on the fracture strength of polymethylmethacrylate. *Mater. Sci. Eng. A* **2004**, *382*, 288–294. [CrossRef]

14. Taylor, D. The theory of critical distances applied to the prediction of brittle fracture in metallic materials. *Struct. Durab. Health Monit.* **2005**, *1*, 145. [CrossRef]

15. Madrazo, V.; Cicero, S.; Carrascal, I.A. On the Point Method and the Line Method notch effect predictions in Al7075-T651. *Eng. Fract. Mech.* **2012**, *79*, 363–379. [CrossRef]

16. Ibáñez-Gutiérrez, F.T.; Cicero, S. Fracture assessment of notched short glass fibre reinforced polyamide 6: An approach from failure assessment diagrams and the theory of critical distances. *Compos. Part B Eng.* **2017**, *111*, 124–133. [CrossRef]

17. Cicero, S.; García, T.; Castro, J.; Madrazo, V.; Andrés, D. Analysis of notch effect on the fracture behaviour of granite and limestone: An approach from the Theory of Critical Distances. *Eng. Geol.* **2014**, *177*, 1–9. [CrossRef]

18. Castro, J.; Cicero, S.; Sagaseta, C. A Criterion for Brittle Failure of Rocks Using the Theory of Critical Distances. *Rock Mech. Rock Eng.* **2015**, *49*, 63–77. [CrossRef]

19. Taylor, D. *The Theory of Critical Distances: A New Perspective in Fracture Mechanics*; Elsevier: London, UK, 2007.

20. Gutiérrez, P.; Cicero, S.; Martínez, B.; Álvarez, J.A. A Theory of Critical Distances based methodology for the analysis of environmentally assisted cracking in steels. *Eng. Fract. Mech.* **2019**, *214*, 134–148.

21. Gallo, P.; Yan, Y.; Sumigawa, T.; Kitamura, T. Fracture Behavior of Nanoscale Notched Silicon Beams Investigated by the Theory of Critical Distances. *Adv. Theory Simul.* **2017**, *1*, 1700006. [CrossRef]

22. Gallo, P.; Sapora, A. Brittle Failure of Nanoscale Notched Silicon Cantilevers: A Finite Fracture Mechanics Approach. *Appl. Sci.* **2020**, *10*, 1640. [CrossRef]
23. Neuber, H. Theorie der technischen Formzahl. *Forsch. Auf. Dem. Geb. Des. Ing. A* **1936**, *7*, 271–274. [CrossRef]
24. Peterson, R.E. *Notch Sensitivity, Metal Fatigue*; Sines, G., Lwaisman, J., Eds.; McGrawrHill: New York, NY, USA, 1959.
25. *EN 573-1:2005, Aluminium and Aluminium Alloys—Chemical Composition and Form of Wrought Products*; European Committee for Standardization: Brussels, Belgium, 2015.
26. *ASTM E8/E8M-16ae1, Standard Test Methods for Tension Testing of Metallic Materials*; ASTM International: West Conshohocken, PA, USA, 2016.
27. *ASTM E1820-20, Standard Test Method for Measurement of Fracture Toughness*; ASTM International: West Conshohocken, PA, USA, 2020.
28. Cicero, S.; Madrazo, V.; García, T.; Cuervo, J.; Ruiz, E. On the notch effect in load bearing capacity, apparent fracture toughness and fracture mechanisms of polymer PMMA, aluminium alloy Al7075-T651 and structural steels S275JR and S355J2. *Eng. Fail. Anal.* **2013**, *29*, 108–121. [CrossRef]
29. Cicero, S.; Madrazo, V.; García, T. On the assessment of U-shaped notches using Failure Assessment Diagrams and the Line Method: Experimental overview and validation. *Theor. Appl. Fract. Mech.* **2015**, *80*, 235–241. [CrossRef]
30. Berto, F.; Lazzarin, P. Recent developments in brittle and quasi-brittle failure assessment of engineering materials by means of local approaches. *Mater. Sci. Eng. R Rep.* **2014**, *75*, 1–48. [CrossRef]

Publisher's Note: MDPI stays neutral with regard to jurisdictional claims in published maps and institutional affiliations.

Article

Fatigue Life and Reliability of Steel Castings through Integrated Simulations and Experiments

Muhammad Azhar Ali Khan [1,*], Anwar Khalil Sheikh [2], Zuhair Mattoug Gasem [2] and Muhammad Asad [1]

[1] Mechanical Engineering Department, College of Engineering, Prince Mohammad Bin Fahd University, Alkhobar 34754, Saudi Arabia; masad@pmu.edu.sa

[2] Mechanical Engineering Department, King Fahd University of Petroleum and Minerals, Dhahran 31261, Saudi Arabia; anwarks1948@gmail.com (A.K.S.); zuhair@kfupm.edu.sa (Z.M.G.)

* Correspondence: mkhan6@pmu.edu.sa

Abstract: The quality and performance of steel castings is always a concern due to porosities formed during solidification of the melt. Nowadays, computational tools are playing a pivotal role in analyzing such defects, followed by their minimization through mold design optimization. Even if the castings are produced with defects in a permissible range, it is important to examine their service life and performance with those defects in a virtual domain using simulation software. This paper aims to develop a methodology with a similar idea of simulation-based optimization of mold design and predictions of life and reliability of components manufactured with minimized casting defects, especially porosities. The cast parts are standard fatigue specimens which are produced through an optimized multi-cavity mold. X-ray imaging is done to determine the soundness of cast parts. Experimental work includes load-controlled fatigue testing under fully reversed condition. The fatigue life of specimens is also simulated and compared with the experimental results. The classical strength-stress model is used to determine the reliability of cast parts through which a safe-load induced stress of steel castings is determined. Finally, probability distributions are fit to the reliability results to develop the reliability models. It is found that porosities can be minimized significantly in the mold design phase using casting simulations. Nevertheless, some porosities are bound to exist, which must be included in realistic estimation of fatigue life and reliability of cast parts.

Keywords: metal casting; mold design; simulation; optimization; fatigue life; reliability

Citation: Khan, M.A.A.; Sheikh, A.K.; Gasem, Z.M.; Asad, M. Fatigue Life and Reliability of Steel Castings through Integrated Simulations and Experiments. *Metals* **2022**, *12*, 339. https://doi.org/10.3390/met12020339

Academic Editors: Ricardo Branco, Filippo Berto and Shengchuan Wu

Received: 29 December 2021
Accepted: 25 January 2022
Published: 15 February 2022

Publisher's Note: MDPI stays neutral with regard to jurisdictional claims in published maps and institutional affiliations.

1. Introduction

Steel casting is widely used for components with high requirements regarding strength, toughness, and wear resistance [1]. Due to a large variety of grades available for steel, it is convenient to adjust the properties such as strength and hardness, corrosion and wear resistance, and operating temperatures. The key advantage of casting steel is obtaining a near net shape product with almost any level of intricacy. However, steel tends to form pores during solidification, which must be considered during the mold design. It is essential to include suitable gate geometries and feeders to minimize shrinkage porosities, as well as a proper venting system for resulting gases, to minimize gas porosities. These porosities often pose challenges in quality, life and reliability of steel castings while in service. Hence, these porosities must be included in realistic fatigue life and reliability estimates of steel castings.

The advancements in computational tools have led to the visualization of both the casting process and performance in a completely virtual domain. In pursuit of a robust mold design, it allows a shift from the conventional trial-and-error approach to a modern proof-of-concept approach which makes the process faster, more efficient and less expensive [2]. The simulation-based optimization is a relatively new idea for developing a robust mold design [3–7]. Demler et al. used casting simulation software to determine suitable casting parameters and to predict porosity in components of a drive train made of low alloy steel [4]. Simulation results suggested a suitable casting temperature of 1680 °C together with a casting time of 10 s. The exact locations of feeders are also identified using simulations to ensure proper filling of the molten metal in critical areas of the casting cavity. Lei et al. studied the optimization of the casting system of turbocharger castings using MAGMASoft [5]. The castings originally produced in the foundry are found with defects such as air entrapment, shrinkage and micro-shrinkage, etc. The causes of such defects are first identified using casting simulations, followed by mold design and process parameter optimization. In this regard, the geometry of the sprue and sprue base is significantly modified, and the pouring time is reduced. This resulted in a steadier flow of the melt within the mold together with a faster filling time. It is reported that the defected casting rate dropped from 20% to 10% using simulation-based modified casting system. Sunanda et al. studied the sand casting optimization of a medium carbon steel pulley using Procast software [6]. The main problem is a very heated central region of the pulley casting during solidification leading to hotspots and shrinkage porosity. The issue is addressed by modifying dimensions of sprue, risers and even the ingates in the casting system. Kumar et al. optimized the gating system of a rotary adapter produced by an investment casting process [7]. The casting process is simulated using Procast. Various parameters are considered such as air, flow length, foreign metal entrapment, in gate velocity and gating ratio. The results for each simulation run are examined in terms of in gate entry velocity, fraction solid, shrinkage porosity, etc. Finally, an optimized gating system is developed, simulated and found free from the defects.

A holistic approach is to include the defects, particularly porosities, predicted in casting simulations while determining the service life and reliability. Efforts have been made in the past towards such integration [8–11]. Sheikh et al. studied the effect of mold design optimization on fatigue life and reliability of cast parts [8]. Casting simulations are used to minimize porosity in cast parts followed by their life prediction and reliability assessment through finite element simulations. A comparison of simulation and experimental results validated the developed methodology and its application to any cast metal/alloy. Schmiedel et al. investigated the fatigue life of cast 42CrMo4 steel in the range from high to very high cycle fatigue [9]. The most detrimental defect in cast state is found to be microshrinkage. A short crack growth model based on considerations of Miller is adapted to examine the fatigue life by using the experimental fatigue data and fracture morphology.

This paper presents an integrated approach of utilizing simulations and experiments to estimate fatigue life and reliability of steel castings. Unlike the conventional method of drawing specimens from already cast plates, rods, blocks or even actual castings, the fatigue specimens are considered to be simple cast parts. A multi-cavity initial mold design is developed, simulated and optimized to produce specimens with minimum porosity. The specimens are cast using an optimized mold design for mechanical testing. Next, fatigue life of specimens is simulated using ABAQUS and the results are compared with experimental results to validate the model. Reliability computations are done using a Stress-Strength model which considers both strength and stress as variables. Finally, probability distributions are fit to the reliability results to develop the reliability models. A graphical representation of the methodology used in this study is presented in Figure 1.

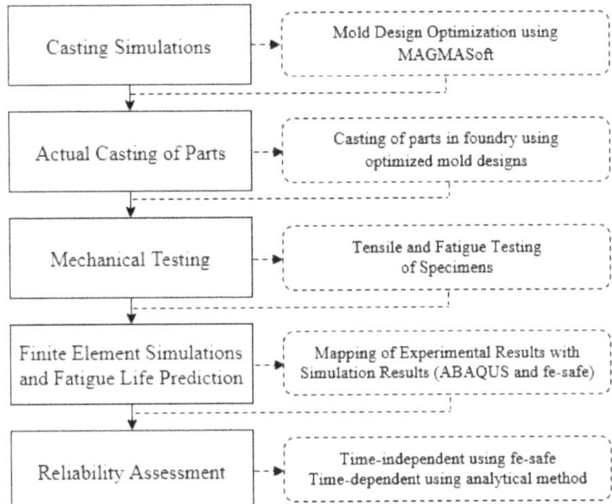

Figure 1. Methodology to estimate fatigue life and reliability of steel cast specimens.

2. Materials

The selected material for this work is ASTM A216 WCB steel due to its widespread use in casting industry. It offers a good mix of strength and ductility, making it suitable to be employed in casting of engineering components. The chemical composition and mechanical properties at room temperature for this steel alloy are provided in Table 1. In addition to the cast material, the selected mold material for this work is Furan sand, which is a chemically bonded sand and does not require baking of the mold. Furan sand molds are compact and provide numerous advantages such as accuracy in casting size, clear outline of casting, smooth surface, good appearance quality, and compact microstructure. Table 2 lists the important properties of the mold material.

Table 1. ASTM A216 WCB steel material specification.

Chemical Composition (Wt.%)									
Fe	C	Mn	Si	P	S	Ni	Cr	Mo	Cu
96.2	0.3	1	0.6	0.035	0.35	0.5	0.5	0.2	0.3

Mechanical Properties (at room temperature as per ASTM A216)		
Yield Strength (MPa)	Tensile Strength (MPa)	Elongation (%)
248	485	22

Table 2. Mold Material Specification.

Mold Material	Furan
Base Materials	Silica Sand
Binder	Furan
Water Content	0%
Initial Temperature	20 °C
Erosion Properties	
Reference Velocity	3 m/s
Reference Time	6 s
Sand Inclusion Parameters	
Sand grain diameter	240 μm
Sand grain density	2650 kg/m^3

3. MAGMASoft Simulations for Casting Specimens

A usual practice in mechanical characterization of metal castings is extracting the specimens from plates, blocks, rods or even actual cast parts. This approach is reasonable when the properties and behavior of actual castings need to be determined under static and dynamic loading. The process-oriented nature of this work demands considering fatigue specimens shown in Figure 2 as a simple cast part. Therefore, a mold is initially designed and simulated to understand filling and solidification behavior and defect predictions, followed by its optimization to minimize porosities.

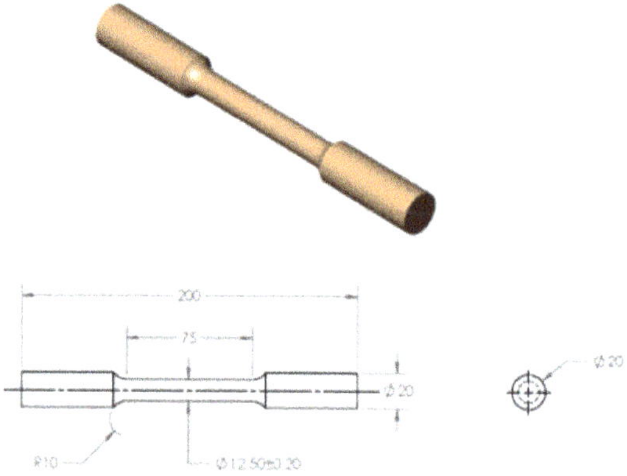

Figure 2. Fatigue specimen geometry (All dimensions are in mm).

Casting simulations are set up in MAGMASoft using different perspectives within the software. To begin with, casting layout, i.e., assembly of components, such as pouring basin, sprue, runner, gates, casting, risers, etc., is either drawn or imported from a CAD software in the *Geometry* perspective. A mold is also created in the same perspective. Next, the casting layout and mold is discretized using *Mesh* perspective. Following mesh generation, material (cast alloy and mold) properties and process parameters are defined using the *Definition* perspective. If needed, *Optimization* perspective can be used to set criteria for autonomous optimization of casting layout. The required results are defined, and simulation settings are done using the *Simulation* perspective. At the end of the simulation, the results are viewed in the *Results* perspective. A flow chart of the simulation sequence in MAGMASoft is presented in Figure 3.

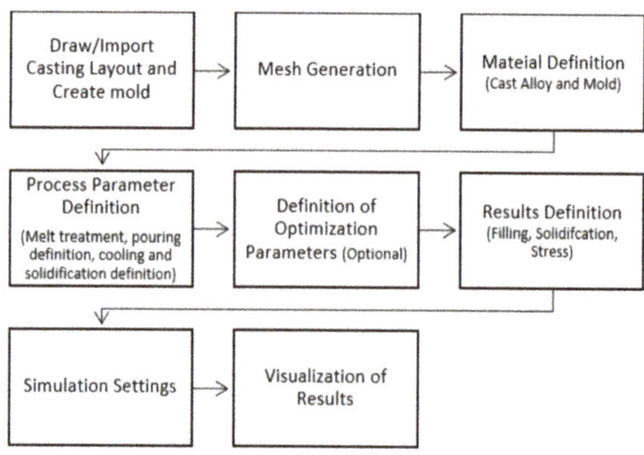

Figure 3. Simulation sequence in MAGMASoft.

The optimized mold designs for casting fatigue specimens are developed as follows:

(a) *Initial Mold Design:* This mold design requires theoretical calculations, casting standards, design expertise of foundrymen and daily foundry practices.

(b) *Casting Simulation of Initial Mold Design:* The resulting mold from (a) is simulated for filling and solidification behavior, stress distribution and porosity prediction in specimens using *MAGMASoft*.

(c) *Optimized Mold Design:* Based on the results obtained in (b), the mold design is carefully optimized for casting layout and/or process parameters. It is important to set the objectives of this optimization, for example, yield maximization, defect minimization, residual stress minimization, etc. In this study, the main objective is set as porosity minimization.

The details of mold design optimization are presented in the following sub-sections.

3.1. Initial Mold Design

The initial mold design is modeled in SOLIDWORKS as shown in Figure 4. It is a multi-cavity mold with sprue-runner configuration, where, instead of choking each casting separately, a single choke is used in the runner area. Casting simulations are done using MAGMASoft. The casting layout is divided into 1,975,320 volume elements using a cubical mesh. MAGMASoft offers a full range of material properties and heat transfer settings as an input to simulations [12]. The pouring temperature is set to be 1630 °C whereas the mold is assumed to be at room temperature, i.e., 20 °C. Pouring time is defined as 15 s. With these simulation settings, the feeding effectivity calculated by the software is ~30%. From a results perspective, it is decided to run simulations for filling and solidification sequence, residual stress distribution, and magnitude and locations of porosities.

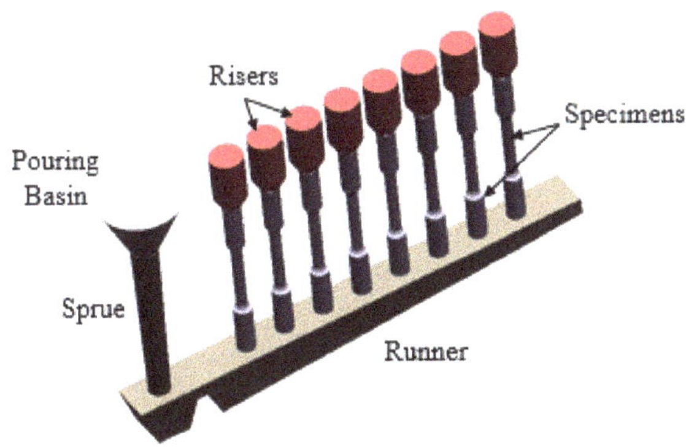

Figure 4. Initial mold design for casting fatigue specimens.

Figure 5a depicts the temperature profile of the mold after pouring. As expected, solidification continues with a drop in temperature of the melt. It can be observed that solidification began with the test section of the specimens and all specimens are solidified at 50% solidification as shown in Figure 5b. The solidification time is recorded to be 8 min with risers being the last region to solidify in the mold. The distribution of residual stresses is presented in Figure 5c where stresses are concentrated in the test section of the specimens. The maximum residual stress at ejection is about 50 MPa, which is a typical magnitude of residual stress in steel castings [13].

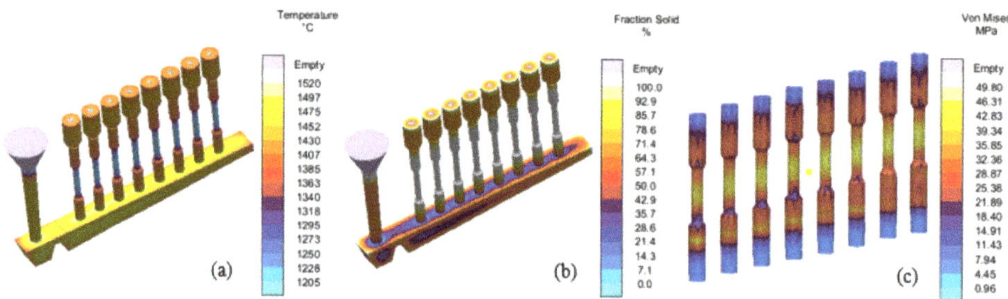

Figure 5. (**a**) Temperature profile within the mold at 50% solidification, (**b**) Percentage fraction solid at 50% solidification, and (**c**) Residual stresses distribution.

In terms of porosity, the X-ray view shown in Figure 6a revealed only one specimen to be pore-free. However, significant porosity is observed in specimens 4, 6, 7 and 8 as shown in Figure 6. Despite the porosity being observed in the grips of test specimens, it had to be minimized for the reason that these specimens are simple cast products. Microporosity is distributed throughout the casting layout with a maximum microporosity to be ~4% as shown in Figure 6b. Figure 6c shows the total porosity in the specimens which is found to be nearly the same in all specimens and could be minimized with a better mold design.

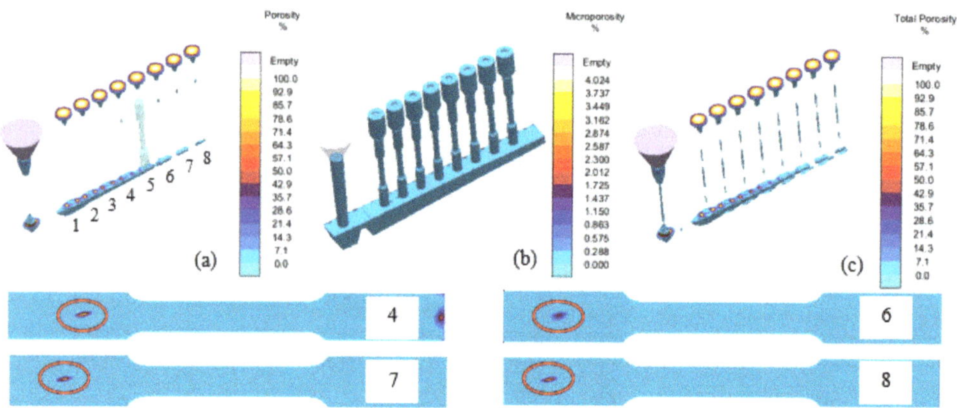

Figure 6. X-ray views of (**a**) porosity, (**b**) microporosity and (**c**) total porosity in simulated cast specimens using initial mold design.

3.2. Optimized Mold Design

The optimization problem is to obtain high quality casting nearly free from defects especially porosity. The objective function is to minimize porosity in castings (test specimens in this case) so that they have a mechanical performance similar to their sound counterparts. The design variables are elements of gating and runner system (Shape and size of sprue, choking, runner geometry and shape, addition of exothermic sleeves, etc.) and the constraints are cast and mold material, melting and pouring temperatures. The new mold design includes changes such as removal of choke from the sprue, choke at the beginning of runner bar, and use of exothermic sleeves. The new casting layout, as shown in Figure 7 is divided into 1,989,414 elements. The simulation settings are kept the same as already discussed earlier. Once again, simulation results indicated temperature gradient in specimens as shown in Figure 8a, and the risers are found to solidify at the end as shown in Figure 8b. It took 16 min for complete solidification using this mold design.

Residual stresses in specimens are shown in Figure 8c which are reduced from ~50 MPa to ~30 MPa. Once again, the X-ray view of the software, as shown in Figure 9, enabled analysis of the simulated porosity, microporosity, and total porosity in specimens. It can be observed that nearly the same porosity is predicted in all specimens as shown in Figure 9a. The connections between runner and specimens are found to be pore-free in the new mold design. Some microporosity is observed, which is uniformly distributed in the specimens as shown in Figure 9b. Total porosity is found to be identical and reduced in all specimens as compared to the total porosity shown in Figure 6c.

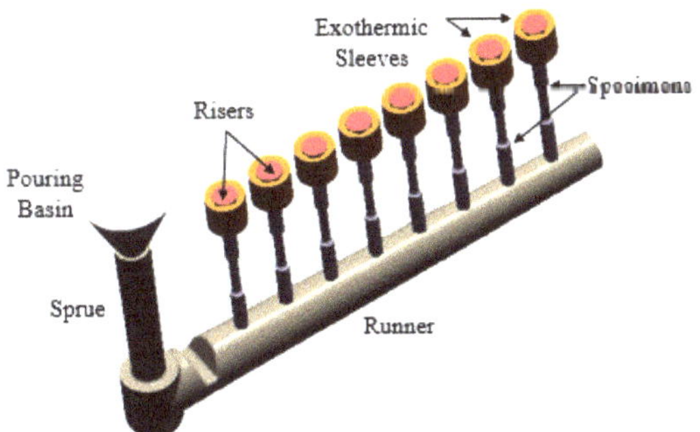

Figure 7. Optimized mold design for casting fatigue specimens.

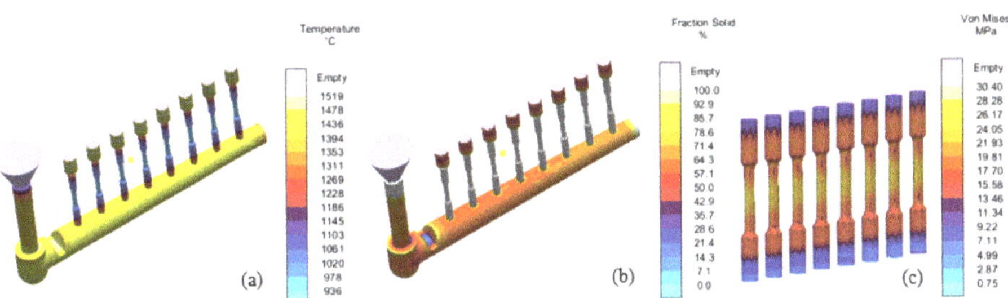

Figure 8. (**a**) Temperature profile within the mold at 50% solidification, (**b**) Percentage fraction solid at 50% solidification, and (**c**) Residual stresses in specimens at ejection.

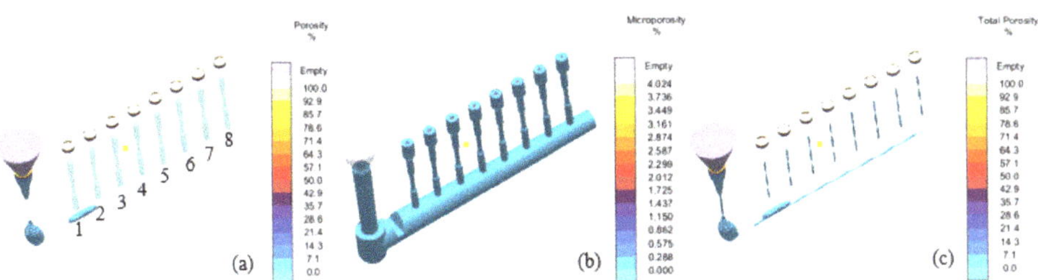

Figure 9. X-ray views of (**a**) porosity, (**b**) microporosity and (**c**) total porosity in simulated cast specimens using optimized mold design.

4. Casting and Mechanical Testing

The preparation of a mold requires a pattern which is an exact replica of the part to be cast. A wooden pattern of the optimized casting layout is developed as shown in Figure 10a to compact sand around it thereby forming a mold. The resulting sand mold is shown in Figure 10b. In order to avoid direct interaction of molten metal with the mold material, a dense layer of zirconium-based coating is applied to the mold surface. Mold properties evaluation revealed a compression strength of 18 kg/cm^2 and a scratch hardness of 28–29. The pouring temperature is set to 1590–1600 °C for which steel is melted in an induction furnace which offers ease of operation, high quality of products, better metallurgical functions, and lower oxidation losses. The steel is deoxidized using Aluminum in the ladle prior to pouring into the mold. It is ensured to perform casting at similar conditions which are used in the simulations. Figure 10c shows the entire casting just after the removal of mold. As mentioned earlier, the specimens are considered as simple cast parts which are obtained after removing pouring basin, sprue, runner bar, and risers. The specimens are then annealed at 920 °C, held for half an hour, and cooled in furnace. Specimens after heat treatment are machined to final dimensions as per the ASTM E-466 standard [14]. The quality and surface finish of the specimens is enhanced during machining.

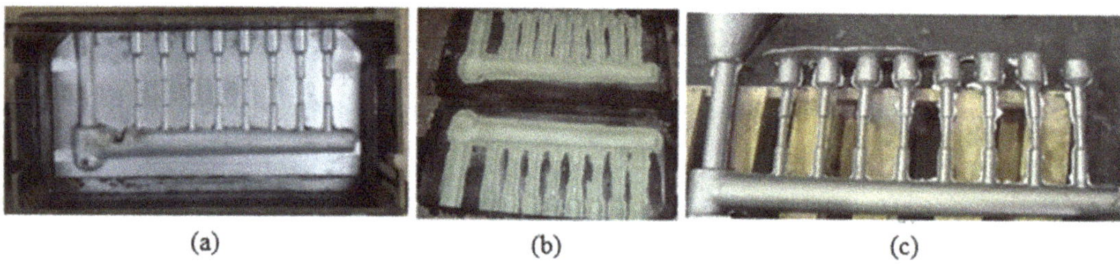

(a)   (b)   (c)

Figure 10. (**a**) Pattern, (**b**) Sand mold and (**c**) Cast Specimens prior to cleaning and finishing.

The quality of cast specimens is evaluated using radiographic examination. The results of X-ray imaging are presented in Figure 11. Some porosity is revealed within the test section of specimens which is observed as difference in appearance of grips and the test section as shown in Figure 11. However, the distribution of porosity from one specimen to another is not much different.

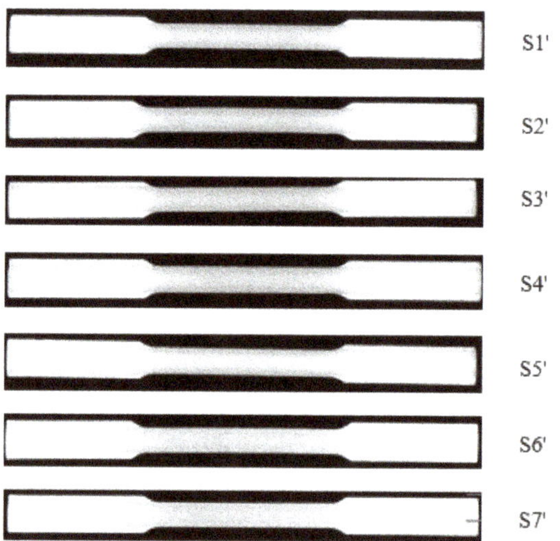

Figure 11. X-ray imaging of steel fatigue specimens.

The fatigue testing is done as per the ASTM E-466 standard [14]. The experimental setup is presented in Figure 12. Fully revered conditions where R = −1 is used to test all specimens under cyclic loading. The frequency of testing is set to be 5 Hz for all specimens. Fatigue testing is done until fracture for all specimens except for runout condition which is 10^6 cycles in this study. A careful selection of stress amplitude has led to the testing of six specimens for finite life and one for the infinite life. Table 3 lists the stress amplitude selected and number of cycles to failure for each specimen. Figure 13 depicts the SN curve developed as a results of fatigue testing of steel specimens.

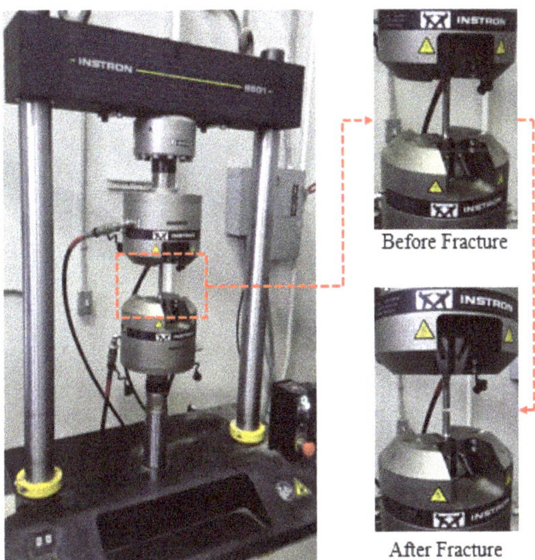

Figure 12. Experimental setup for fatigue testing.

Table 3. Experimental results for fatigue testing of specimens.

Specimen ID	Stress Applied for Test (MPa)	Frequency (Hz)	Fatigue Life (Cycles)
S1′	326	5	62
S2′	244	5	6545
S3′	204	5	20,561
S4′	163	5	56,893
S5′	122	5	145,089
S6′	81	5	535,564
S7′	41	5	3,061,195

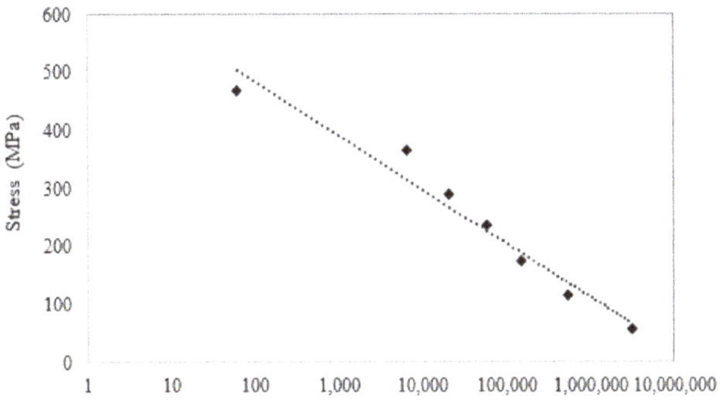

Figure 13. S-N curve for cast steel specimens.

5. Fatigue Simulations and Life Prediction

5.1. Simulation Procedure

Fatigue testing is simulated by doing (a) finite element elastic stress using the load in experimental fatigue testing and (b) fatigue life prediction using multi-axial strain-life approach using stress fields predicted in (a). The results of porosity from MAGMASoft are mapped to Finite Element Analysis (FEA) nodes using MAGMAlink. MAGMAlink allows user to import and export results to and from other softwares into and out of MAGMASoft. The translation and rotation features enable FEA mesh to accurately overlay the MAGMASoft model. FEA mesh is developed in ABAQUS prior to stress simulation. MAGMAlink provides the magnitude of nodal porosity which has to be integrated in ABAQUS. The node sets and nodal porosity data are included in the ABAQUS input file which contains all commands, boundary conditions, and properties required to run the stress simulation. A comparison of experimental and simulation results ensures competency of the model in predicting the fatigue life prediction.

The element type used for fatigue simulations is an eight-node linear brick element (C3D8R). The boundary conditions are set to replicate the actual testing conditions. The specimen is held fixed from one end and a uniformly distributed load, reflecting the actual loading condition for each specimen, is applied to the other end as shown in Figure 14a. Mesh sensitivity analysis confirmed a 1 mm node spacing suitable for all simulations. The resulting mesh is shown in Figure 14b. The finite element model developed with these specifications consist of 83,433 elements, 90,373 nodes and 345,546 variables.

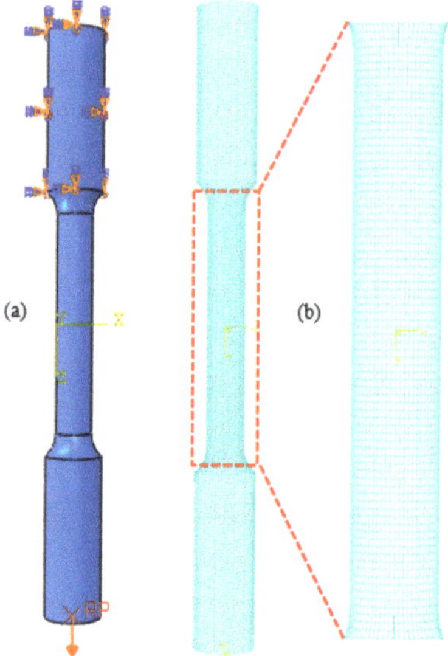

Figure 14. (a) Boundary conditions and (b) Meshed specimen with 1 mm node spacing.

The simulation results are imported to Fe-safe software [15] which is used to predict lives of specimens. Tension and compression steps of fully reversed loading, i.e., R = 1 are taken into consideration while calculating the fatigue life. The material properties used for life prediction are summarized in Table 4. The inputs required by the software are material properties and loading cycle. Stress-strain conversion takes place within Fe-safe by utilizing the elastic modulus of pore-free steel. The recommended algorithm by Fe-safe for predicting fatigue life of steel is Brown–Miller algorithm with Morrow mean stress correction [15].

Table 4. ASTM A216 WCB Steel Monotonic and Cyclic Properties [1,16].

ASTM A216 WCB Monotonic Properties		
Ultimate Tensile Strength, σ_{UTS}	(MPa)	620
Yield Strength, σ_y	(MPa)	355
Modulus of Elasticity, E_0	(MPa)	198,000
Percentage Elongation, %EL	-	32.5
Fracture strength, σ_f	(MPa)	293
ASTM A216 WCB Cyclic Properties		
Fatigue Strength, S_f	(MPa)	141
S_f/σ_{UTS}		0.227
Cyclic Strain Hardening Coefficient, K'	(MPa)	1558
Cyclic Strain Hardening Exponent, n'		0.295
Fatigue Strength Coefficient, σ_f'	(MPa)	1015
Fatigue Strength Exponent, b		−0.136
Fatigue Ductility Coefficient, ε_f'		0.195
Fatigue Ductility Exponent, c		−0.435

5.2. Brown–Miller Analysis

The Brown–Miller algorithm conservative approach for fatigue life prediction, using planes perpendicular to the surface and at 45 degrees to the surface. It uses a critical plane analysis to estimate the fatigue life in reversals to failure, $2N_f$, by solving the following equation [15] at each node.

$$\frac{\Delta\gamma_{max}}{2} + \frac{\Delta\varepsilon_n}{2} = 1.65\frac{\sigma'_f}{E}(2N_f)^b + 1.75\ \varepsilon'_f(2N_f)^c \tag{1}$$

With Morrow mean stress correction, Equation (1) is modified to

$$\frac{\Delta\gamma_{max}}{2} + \frac{\Delta\varepsilon_n}{2} = 1.65\frac{(\sigma'_f - \sigma_m)}{E}(2N_f)^b + 1.75\ \varepsilon'_f(2N_f)^c \tag{2}$$

where $\frac{\Delta\gamma_{max}}{2}$ is the maximum shear strain amplitude, $\frac{\Delta\varepsilon_n}{2}$ is the strain amplitude normal to the shear stress plane, σ_m is the mean stress, σ'_f is the fatigue strength coefficient, b is the fatigue strength exponent, ε'_f is the fatigue ductility coefficient, and c is the fatigue ductility exponent.

The critical plane analysis is used to compute the strain tensor at a FE node having three direct and three shear components. The strain tensor is then resolved onto a number of planes, where, at each place the damage associated with the strain is evaluated. The plane resulted with maximum damage is used in strain-life computations. For a Cartesian $x - y - z$ coordinate system, the unique planes can be defined by the orientation the normal of the plane surface makes with respect to the coordinate system [17]. This orientation can be defined by an angle from x-axis toward the y-axis, and another angle from the z-axis toward the $x - y$ plane [15]. Fe-safe searched for the critical plane having worst damage in 10-degree increments over the 180-degree range of the first angle and 90-degree range of the second angle. The strains are projected to the calculation plane using direction cosines.

5.3. Results of Fatigue Life Simulations

Figure 15 shows the simulated stress field in porous specimens, S1′ through S7′, using ABAQUS and the fatigue life predicted by Fe-safe. The scale used for fatigue life is Log10 life, i.e., 1 million cycles are represented by 6 on this scale. As mentioned earlier, a node spacing of 1 mm is found to show good agreement between the measured and the predicted fatigue lives. A complex stress field is developed due to porosity upon loading each specimen. It is observed that area with stress concentrations experienced the shortest lives. Moreover, for each specimen, the least number of cycles to failure is predicted on the surface. This shortest life predicted is considered to be the fatigue life for that specimen. A summary of simulated stress, measured and predicted fatigue life for each specimens, is presented in Table 5.

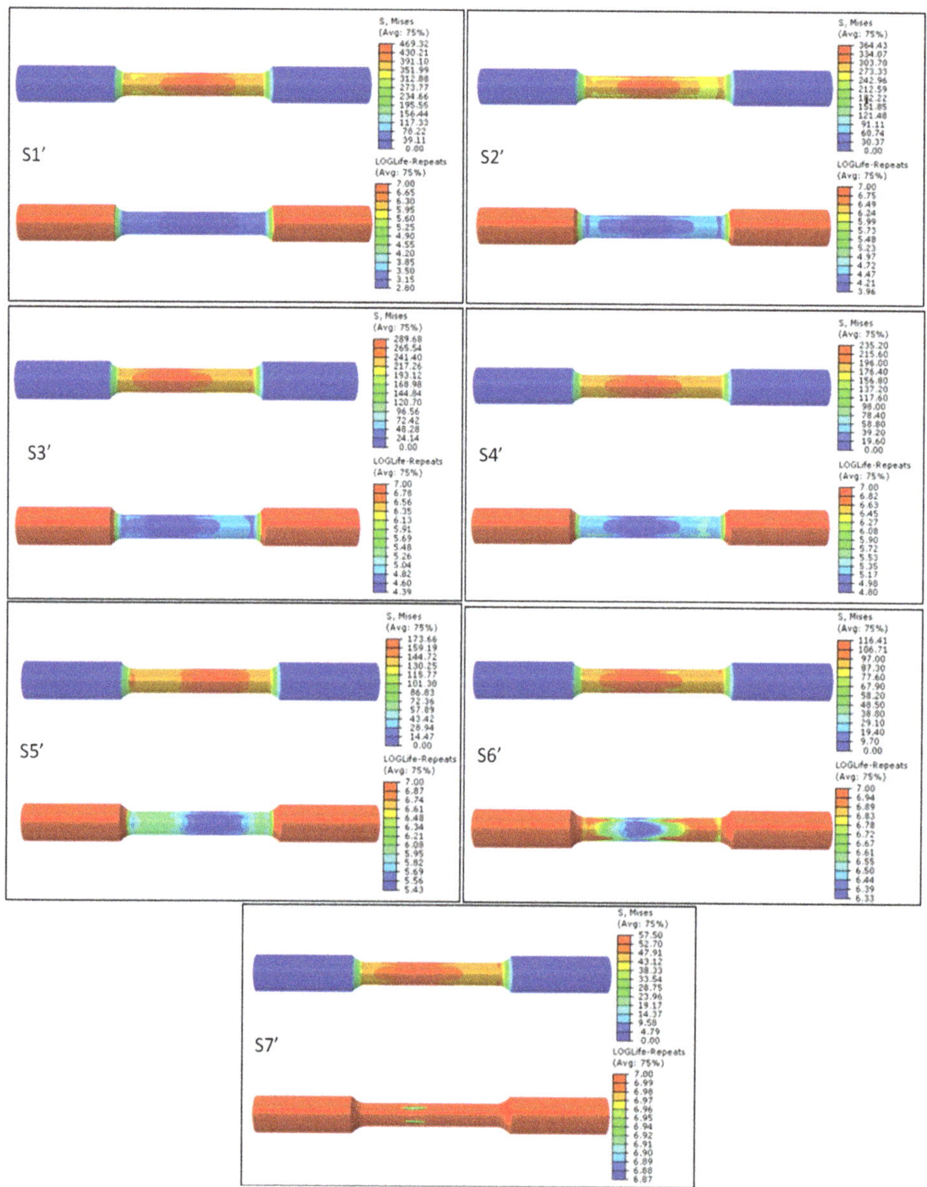

Figure 15. ABAQUS stress analysis and fatigue life prediction from Fe-safe.

Table 5. Summary of simulated and experimental results of fatigue testing for steel specimens.

Specimen ID	Simulated Stress (MPa)	Measured Fatigue Life (Cycles)	Simulated Fatigue Life (Cycle)
S1′	469.32	62	630
S2′	364.43	6545	9115
S3′	289.68	20,561	24,219
S4′	235.20	56,893	62,983
S5′	173.66	145,089	259,682
S6′	116.41	535,564	2,133,921
S7′	57.50	3,061,195	7,421,449

The experimental and simulated S-N curve are presented in Figure 16. Simulation results shows two specimens to experience the infinite life where the predicted number of cycles to failure are beyond the runout condition, i.e., 10^6 cycles. For a better understanding of experimental and simulated results, the measured and simulated lives for all specimens are compared in Figure 17. A line of perfect correspondence is provided in the figure to determine if a prediction is non-conservative (above the line) or conservative (below the line). It can be observed that six out of seven specimens are within a factor of 10 of the test results, which can be considered as a good agreement in fatigue life prediction [18]. The specimen S1′ is found to have a relatively bad agreement since the predicted fatigue life is slightly more than the factor of 10 of the test result. In general, the overall results for fatigue life prediction are encouraging, however, the overall nature of predictions are found to be non-conservative in steel specimens.

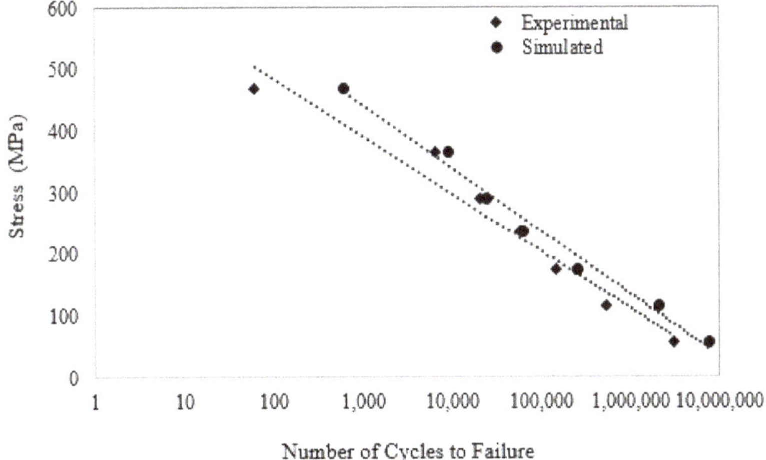

Figure 16. Simulated and experimental S-N curve for steel specimens.

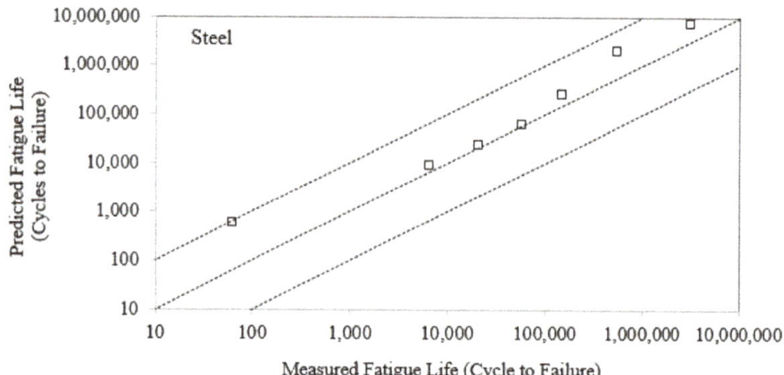

Figure 17. Measured and predicted fatigue lives of steel specimens.

6. Reliability Analysis

6.1. Strength-Stress Model and Reliability Calculations

The reliability of an engineering product deals with the undesirable events or failures during its service life. It can be precisely defined as the reliability of a part is the probability that, when operated under defined set of conditions, the part will perform its intended function adequately for a specified interval of time [19]. It is an established fact that apparently identical parts operating under similar conditions fail at different points in time. This brings about a need to describe failure phenomena in probabilistic terms and therefore, fundamental aspects of reliability heavily rely on concepts from probability.

The classic strength-limited design suggests the strength should be greater than the stress. A design factor is always added to cover the uncertainties. If strength and stress distributions are known, the reliability of a part can be determined using interference theory presented in [8]. For a strength-limited design, let the density function for the strength is f_1 and that for stress is f_2, the reliability function will be a joint probability function, where

$$P(S > \sigma) = P[S - \sigma > 0] = R$$
$$R = \int_{-\infty}^{\infty} f_1 (S) \left[\int_{S}^{\infty} f_2 (\sigma)d\sigma \right] dS \tag{3}$$

where, S is the significant strength and σ is the significant load-induced stress. The task for a given design is to ensure that $S > \sigma$. Based on fatigue life of specimens obtained at different stress levels in Section 5, the reliability of cast specimens is estimated using this model in this work.

Reliability computations are done for two scenarios: (i) Time-independent load-induced stress and (ii) Time-dependent load-induced stress. Four different load-induced stress values are selected based on the expected loading conditions on steel castings, i.e., 79 MPa, 87 Mpa, 96 Mpa and 104 Mpa. FE-safe combines the variability in both material fatigue strength and applied loading (if any), to calculate the probability of failure for a specified life. For time-independent load-induced stress case, the reliability computations are based on normally distributed stress and Weibull distributed strength. The details of failure rate calculations are presented in [8].

The reliability analysis for time-dependent load-induced stress provides a more conservative estimate of component performance during service life. The strength-stress interference theory is also applicable for this scenario, but load-induced stress cannot be modeled through normal distribution. For this reason, Fe-safe could not be used for reliability computations in this case. Instead, analytical methods proposed by Samar et al. [20]

are used, which models both the strength and the stress through Weibull distribution. The probability density function of strength S and stress σ distributions are given by:

$$f_1(S) = \frac{\beta_S}{\theta_S}\left(\frac{S}{\theta_S}\right)^{\beta_S-1}.\exp\left(-\frac{S}{\theta_S}\right)^{\beta_S} \tag{4}$$

$$f_2(\sigma) = \frac{\beta_\sigma}{\theta_\sigma}\left(\frac{\sigma}{\theta_\sigma}\right)^{\beta_\sigma-1}.\exp\left(-\frac{\sigma}{\theta_\sigma}\right)^{\beta_\sigma} \tag{5}$$

And the resultant reliability function is similar to Equation (3). The change in load-induced stress with time can be modeled through Rayleigh distribution, which is a special case of Weibull distribution with shape parameter β equal to 2. If $\beta_S = 2\beta_\sigma$, then the reliability analysis is based on Weibull distributed strength and Rayleigh distributed load-induced stress [8]. Using the results presented by Samar et al. [20] and with $\beta_S = 2\beta_\sigma$, the reliability function is given by

$$R = P(S > \sigma) = \frac{\theta_S}{\theta_\sigma}\sqrt{\pi}.\exp\left(\frac{1}{4}\left(\frac{\theta_S}{\theta_\sigma}\right)^2\right).\left\{1 - \phi\left[\frac{1}{\sqrt{2}}.\left(\frac{\theta_S}{\theta_\sigma}\right)\right]\right\} \tag{6}$$

Hence, the reliability can be estimated against the ratio of scale parameters, i.e., $\frac{\theta_S}{\theta_\sigma}$ for the targeted lives. Here, the $\frac{\theta_S}{\theta_\sigma}$ ratio is approximated to be similar to that of the $\frac{S}{\sigma}$ ratios for the targeted lives.

6.2. Reliability Results

Figures 18–21 depicts the reliability curves for steel specimens with load-induced stress values 79 MPa, 87 MPa, 96 MPa and 104 MPa, respectively. In each of these figures, the reliability of the cast parts is presented with a 0% and 5% variability in the load-induced stress. It can be observed that a 5% variability in load has not significantly affected the component's reliability. This could be due to the fact that the software derives an equivalent loading for non-constant amplitude loadings. However, the component reliability for the same targeted life decreases with increasing load induced stress. Moreover, the effect of Weibull shape parameter β is also analyzed and presented. Previously, it is reported that the coefficient of variation in hardness and strength of the specimens is very less [1], which suggests a higher value of β to be more realistic. Therefore, reliability is computed at various values of β, i.e., 3–5 and 10. The higher the value of β, the components will be more reliable for the same targeted lives as shown in Figures 18–21.

A summary of reliability computations for steel is presented in Figure 22. The plot indicates the reliability of components for the runout conditions used in fatigue life prediction, i.e., 10^6 cycles against the load induced stress. A region of safe loading is defined based on how many components survive at a particular load. It is noted that, independent of β, more than 86% components survive for the infinite life at a load-induced stress of 85 MPa. However, this is a conservative estimate of safe loading on component to allow for possible variations in component strength, which is represented by β in reliability calculations. From experience, it is readily accepted that apparently same components fail at different points of time during service life. Therefore, in a strength-limited design, it is appropriate to consider such variations in reliability computations. Nevertheless, if such variations are assured to be at a minimum, the use of a higher value of β is more realistic, which in this case resulted in a reliability of more than 95% at a load induced stress of 95 MPa in Figure 22. Hence, with the optimized mold design and a higher $\beta = 10$, it is reasonable to infer a safe load-induced stress up to 95 Mpa.

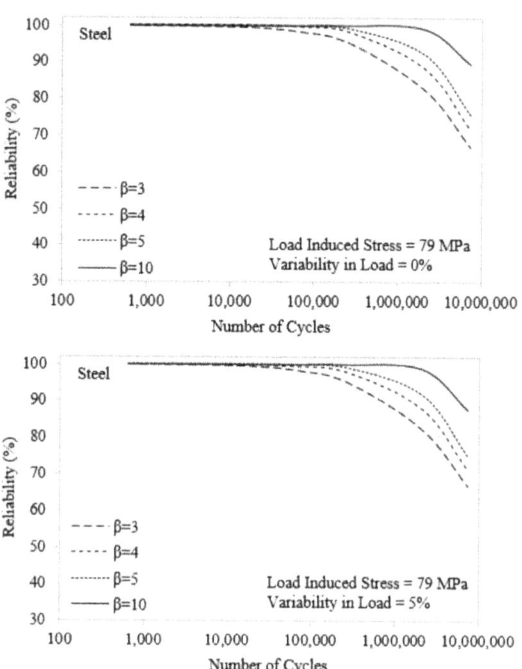

Figure 18. Reliability results with a load-induced stress of 79 MPa.

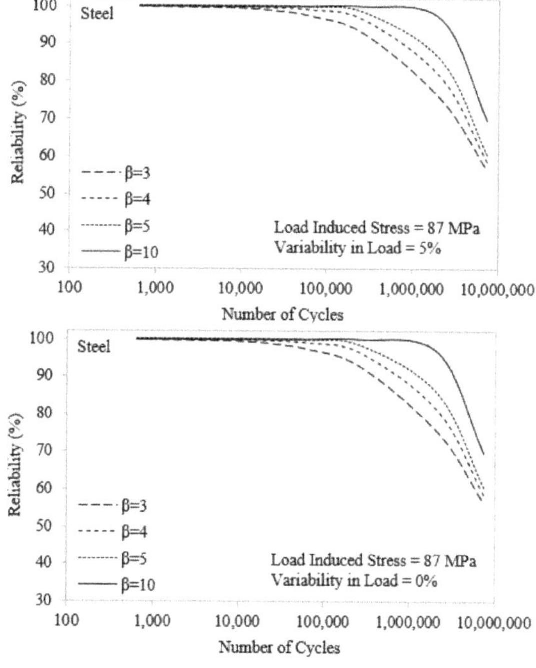

Figure 19. Reliability results with a load-induced stress of 87 MPa.

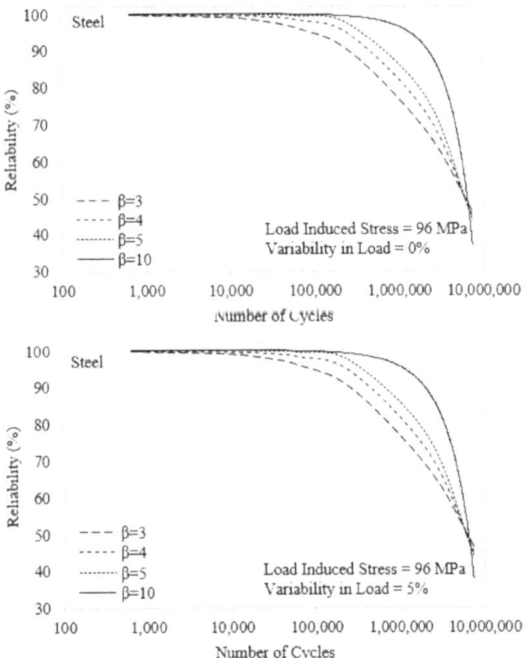

Figure 20. Reliability results with a load-induced stress of 96 MPa.

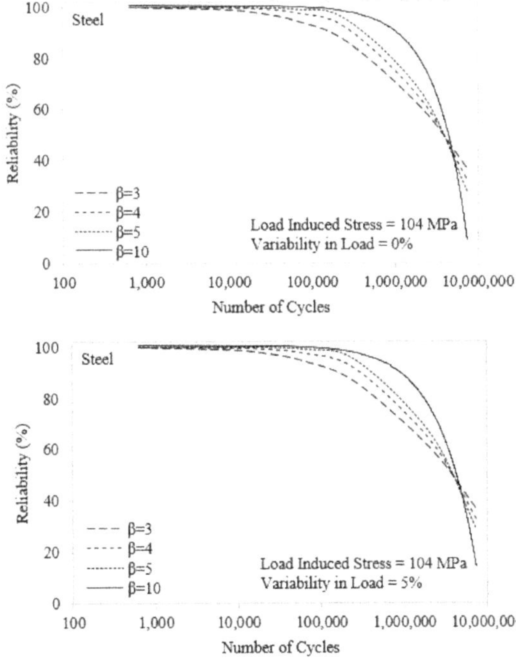

Figure 21. Reliability results with a load-induced stress of 104 MPa.

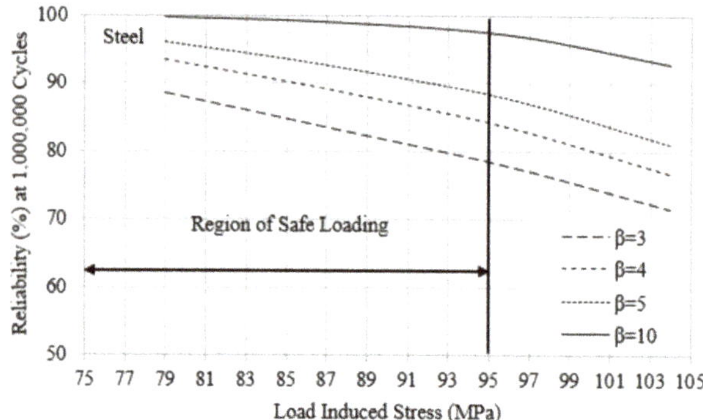

Figure 22. Summary of reliability results for cast steel.

Figure 23 shows the reliability results for time-dependent load induced stress based on Equation (6). Here, a plot of reliability of component versus the ratio of scale parameters, i.e., $\frac{\theta_S}{\theta_\sigma}$ in Weibull distribution is made to evaluate the results. It is evident from Figure 23 that the reliability of a component increases with increasing design factor. If the material's strength S is four times the mean load-induced stress σ, the parts result in ~90% reliability which reduces to 54.5% when $S = \sigma$. It should be noted that the results in Figure 23 are valid for $\beta_S = 2\beta_\sigma$ which suggests that the strength is Weibull distributed and load-induced stress is Rayleigh distributed.

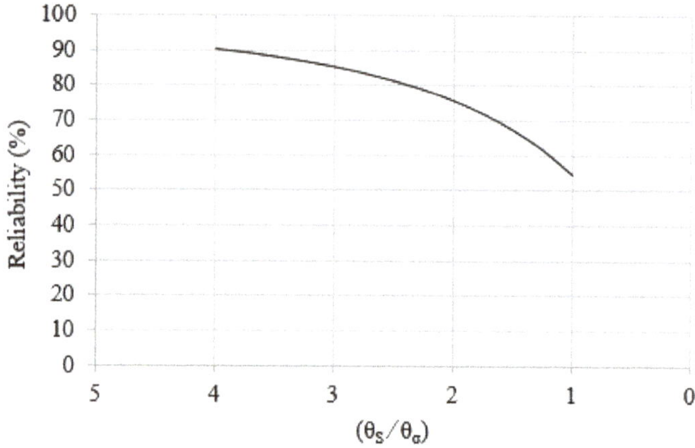

Figure 23. Reliability results for time-dependent load-induced stress.

6.3. Distribution Fitting to Reliability Results

The reliability results presented above can be used to determine reliability models by fitting probability distributions to these estimates. In this regard, log-normal distribution and Weibull distribution for the 79 MPa stress with 5% variability in load are used. The distribution fitting includes linearizing the reliability function, plotting of linearized reliability function, and finally estimating the distribution parameters. The log-normal distribution parameters are μ and σ, whereas, the Weibull distribution parameters are β and θ. For brevity, the procedure to obtain fitted models is introduced here briefly and the details are presented elsewhere [8]. The results for log-normal distribution and Weibull distribution fitted to reliability estimates are presented in Figures 24 and 25, respectively.

The equations of fitted models are also included to determine the distribution parameters, which are summarized in Tables 6 and 7.

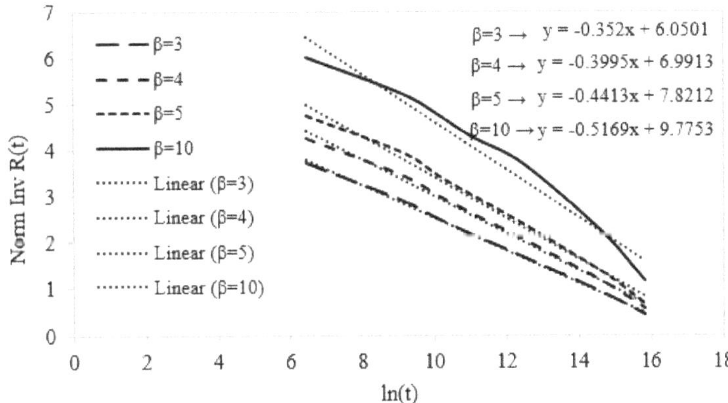

Figure 24. Log-Normal distribution fitted to reliability estimates with load-induced stress 79 MPa.

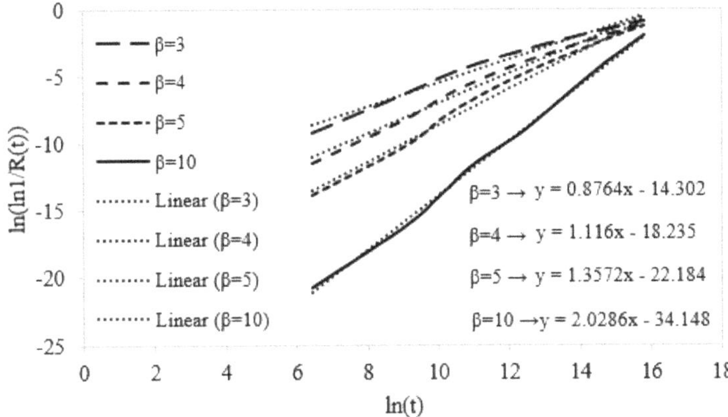

Figure 25. Weibull distribution fitted to reliability estimates with load-induced stress 79 MPa.

Table 6. Log-normal distribution parameters fitted to reliability estimates.

Weibull Shape Parameter for Reliability Estimate	Log-Normal Distribution Parameters	
	μ	σ
β = 3	17.19	2.84
β = 4	17.50	2.50
β = 5	17.72	2.27
β = 10	18.91	1.93

Table 7. Weibull distribution parameters fitted to reliability estimates.

Weibull Shape Parameter for Reliability Estimate	Weibull Distribution Parameters	
	β	θ
$\beta = 3$	0.876	15,464,922
$\beta = 4$	1.116	13,564,126
$\beta = 5$	1.357	12,178,806
$\beta = 10$	2.082	100,931,406

7. Conclusions and Future Work

This study presents a method for determining the fatigue life and reliability of ASTM A 216 WCB steel castings using advanced simulations tools. For simplicity, standard fatigue specimens are produced with an optimized and robust mold design. The life and reliability are predicted taking into consideration the porosities, which are minimized, yet exist to some extent in the specimens produced with optimized mold design. The key conclusions are as follows:

- Casting simulation softwares such as MAGMASoft are capable of examining the effects of several factors such as temperature of molten metal, pouring time and velocity, gating and runner design, riser design, and mold configurations on the quality of castings.
- Testing of cast standard specimens is a practical approach to validate the quality of castings produced using simulation-based optimized mold designs.
- The simulations done in this work utilizes the porosity fractions which are defined over a volume that is large compared to microscopic pore geometry. The good agreement between measured and simulated fatigue lives confirms the adequate mesh used in life prediction with integrated porosity. Nevertheless, it can be expected that stress concentration around very small shrinkage pores can be modeled better with further mesh refinement, which consequently will require longer simulation time and more powerful computational facilities.
- The classical strength-stress interference theory provides realistic estimates reliability. It is concluded that the reliability for infinite life drops down with increasing load-induced stress. The overall results suggest 95 MPa as a safe stress to survive the cast steel parts for an infinite life.
- The methodology presented in this work can be applied to real cast parts. Besides cast steels, the methodology is deemed robust in cradle to grave analysis of cast parts produced with almost any alloy.

Author Contributions: Conceptualization, M.A.A.K. and A.K.S.; methodology, M.A.A.K.; software, M.A.A.K.; validation, M.A.A.K., Z.M.G. and M.A.; formal analysis, M.A.A.K.; resources, Z.M.G.; data curation, M.A.A.K.; writing—original draft preparation, M.A.A.K. and A.K.S.; writing—review and editing, Z.M.G. and M.A.; supervision, A.K.S. All authors have read and agreed to the published version of the manuscript.

Funding: This research was funded by The National Science, Technology and Innovation Plan (NSTIP), Saudi Arabia, grant number 14-ADV890-04-R and the APC was funded by Prince Moham-mad Bin Fahd University, Saudi Arabia.

Informed Consent Statement: Not Applicable.

Data Availability Statement: Not Applicable.

Acknowledgments: The simulations presented in this study are done at Rapid Prototyping and Reverse Engineering Lab at King Fahd University of Petroleum and Minerals (KFUPM). Special thanks to MAGMA and KFUPM for their continuous support during this work. The authors would also like to acknowledge the supportive environment provided by Prince Mohammad Bin Fahd University to facilitate the preparation of manuscript and its publication.

Conflicts of Interest: The authors declare no conflict of interest.

References

1. Sheikh, A.K.; Khan, M.A.A. Mold Design Optimization and Quality Assessment of Steel Castings through Integrated Simulations and Experiments. *J. Mech. Sci. Technol.* **2020**, *34*, 2975–2983. [CrossRef]
2. Khan, M.A.A.; Sheikh, A.K. A Comparative Study of Simulation Software for Modelling Metal Casting Processes. *Int. J. Simul. Model.* **2018**, *17*, 197–209. [CrossRef]
3. Futáš, P.; Pribulová, A.; Fedorko, G.; Molnár, V.; Junáková, A.; Laskovský, V. Failure Analysis of a Railway Brake Disc with the Use of Casting Process Simulation. *Eng. Fail. Anal.* **2019**, *95*, 226–238. [CrossRef]
4. Demler, E.; Götze, S.; Herbst, S.; Nürnberger, F.; Maier, H.J.; Ursinus, J.; Büdenbender, C.; Behrens, B.-A. Casting Manufacturing of Cylindrical Preforms Made of Low Alloy Steels. *Procedia Manuf.* **2020**, *47*, 445–449. [CrossRef]
5. Lei, C.; Yang, Y.; Yang, G.; Huang, Y. Magma Software Simulation Assisted Optimization of the Casting System of Turbocharger Castings. *Procedia Manuf.* **2019**, *37*, 59–65. [CrossRef]
6. Sunanda, A.; Jagannadha Raju, M.V. Simulation for Prediction Analysis of Defects in Pulley Casted Using Sand Casting Process. *Mater. Today Proc.* **2021**; *in press.* [CrossRef]
7. Kumar, R.; Madhu, S.; Aravindh, K.; Jayakumar, V.; Bharathiraja, G.; Muniappan, A. Casting Design and Simulation of Gating System in Rotary Adaptor Using Procast Software for Defect Minimization. *Mater. Today Proc.* **2020**, *22*, 799–805. [CrossRef]
8. Sheikh, A.K.; Khan, M.A.A. Fatigue Life Prediction and Reliability Assessment of Ductile Iron Castings Using Optimized Mold Design. *Int. J. Adv. Manuf. Technol.* **2020**, *106*, 1945–1966. [CrossRef]
9. Schmiedel, A.; Kirste, T.; Morgenstern, R.; Weidner, A.; Biermann, H. The Fatigue Life of 42CrMo4 Steel in the Range of HCF to VHCF at Elevated Temperatures up to 773 K. *Int. J. Fatigue* **2021**, *152*, 106437. [CrossRef]
10. Foglio, E.; Gelfi, M.; Pola, A.; Goffelli, S.; Lusuardi, D. Fatigue Characterization and Optimization of the Production Process of Heavy Section Ductile Iron Castings. *Int. J. Met.* **2017**, *11*, 33–43. [CrossRef]
11. Borsato, T.; Ferro, P.; Berto, F.; Carollo, C. Mechanical and Fatigue Properties of Pearlitic Ductile Iron Castings Characterized by Long Solidification Times. *Eng. Fail. Anal.* **2017**, *79*, 902–912. [CrossRef]
12. Hattel, J. *Fundamentals of Numerical Modelling of Casting Processes*; Polyteknisk Forlag: Kongens Lyngby, Denmark, 2005; ISBN 978-87-502-0969-0. Available online: https://orbit.dtu.dk/en/publications/fundamentals-of-numerical-modelling-of-casting-processes (accessed on 12 November 2021).
13. Egner-Walter, A.; Kothen, M. Using Stress Simulation to Tackle Distortion and Cracking in Castings. *Metall. Sci. Tecnol.* **2013**, *24*. Available online: http://78.47.113.247/index.php/MST/article/view/1122 (accessed on 12 November 2021).
14. *ASTM E466-15 Standard Practice for Conducting Force Controlled Constant Amplitude Axial Fatigue Tests of Metallic Materials.* 2015. Available online: https://ci.nii.ac.jp/naid/10017283294/ (accessed on 12 November 2021).
15. *Dassault Systems UK Limited Fe-Safe User Manual*, 2015.
16. Khan, M.A.A. *Development of a Simulation-Based Methodology for Mold Design Optimization and Reliability Assessment of Cast Parts-KFUPM EPrints*; King Fahd University of Petroleum and Minerals: Dhahran, Saudi Arabia, 2018.
17. Hardin, R.A.; Beckermann, C. Prediction of the Fatigue Life of Cast Steel Containing Shrinkage Porosity. *Metall. Mater. Trans. A* **2009**, *40*, 581–597. [CrossRef]
18. Hardin, R.A.; Beckermann, C. Integrated Design of Castings: Effect of Porosity on Mechanical Performance. *IOP Conf. Ser. Mater. Sci. Eng.* **2012**, *33*, 012069. [CrossRef]
19. Kapur, K.C.; Lamberson, L.R. *Reliability in Engineering Design*; Wiley: Hoboken, NJ, USA, 1977; ISBN 978-0-471-51191-5. Available online: https://ui.adsabs.harvard.edu/abs/1977nyjw.book.....K/abstract (accessed on 12 November 2021).
20. Samar Ali, S.; Kannan, S. A Diagnostic Approach to Weibull-Weibull Stress-strength Model and Its Generalization. *Int. J. Qual. Reliab. Manag.* **2011**, *28*, 451–463. [CrossRef]

 *metals*

Article

Improving the Fatigue of Newly Designed Mechanical System Subjected to Repeated Impact Loading

Seongwoo Woo [1,*], Dennis L. O'Neal [2], Dereje Engida Woldemichael [1], Samson Mekbib Atnaw [1] and Muluneh Mekonnen Tulu [1]

[1] Department of Mechanical Engineering, College of Electrical and Mechanical Engineering, Addis Ababa Science & Technology University, PO Box 16417 Addis Ababa, Ethiopia; presidentoffice@aastu.edu.et (D.E.W.); samson.mekbib@aastu.edu.et (S.M.A.); muluneh.mekonnen@aastu.edu.et (M.M.T.)

[2] Department of Mechanical Engineering, Dean of Engineering and Computer Science, Baylor University, Waco, TX 76798-7356, USA; dennis_oneal@baylor.edu

* Correspondence: twinwoo@yahoo.com; Tel.: +251-90-047-6711

Abstract: This paper develops parametric accelerated life testing (ALT) as a systematic reliability method to produce the reliability quantitative (RQ) specifications—mission cycle—for recognizing missing design defects in mechanical products as applying the accelerated load, expressed as the inverse of stress ratio, R. Parametric ALT is a way to enhance the prediction of fatigue failure for mechanical systems subjected to repeated impact loading. It incorporates: (1) A parametric ALT plan formed on the system BX lifetime, (2) a fatigue failure and design, (3) customized ALTs with design alternatives, and (4) an assessment of whether the last design(s) of the system fulfills the objective BX lifetime. A BX life concept with a generalized life-stress model and a sample size equation are suggested. A domestic refrigerator hinge kit system (HKS), which was a newly designed mechanical product, was used to illustrate the methodology. The HKS was subjected to repeated impact loading resulting in failure of the HKS in the field. To conduct ALTs, a force and momentum balance was utilized on the HKS. A straightforward impact loading of the HKS in closing the refrigerator door was examined. At the first ALT, the housing of the HKS failed. As an action plan, the hinge kit housing was modified by attaching inside supporting ribs to the HKS to provide sufficient mechanical strength against its loading. At the second ALT, the torsional shaft in the HKS made with austenitic ductile iron (18 wt% Ni) failed. The cracked torsional shaft for the 2nd ALTs came from its insufficient rounding, which failed due to repeated stress. As an action plan, to have sufficient material strength for the repetitive impact loads, the torsional shaft was reshaped to give it more rounding from R0.5 mm to R2.0 mm. After these modifications, there were no problems at the third ALT. The lifetime of the HKS in the domestic refrigerator was assured to be B1 life 10 years.

Keywords: fatigue failure; design flaws; mechanical system; parametric ALT; hinge kit system

Citation: Woo, S.; O'Neal, D.L.; Woldemichael, D.E.; Atnaw, S.M.; Tulu, M.M. Improving the Fatigue of Newly Designed Mechanical System Subjected to Repeated Impact Loading. *Metals* **2021**, *11*, 139. https://doi.org/10.3390/met11010139

Received: 16 December 2020
Accepted: 4 January 2021
Published: 12 January 2021

Publisher's Note: MDPI stays neutral with regard to jurisdictional claims in published maps and institutional affiliations.

1. Introduction

Because of the competitiveness in the global market, manufacturers must continually innovate and improve their products. Often, this involves new technologies and design features for the product that must be quickly delivered to the marketplace. However, without sufficient testing or anticipation of how the features may be used or misused, the introduction of these new features may increase failures of the product in the field and negatively impact the company's image. These added attributes are often requested or desired by consumers, and companies strive to include these features in new design specifications for the product. The features for the newly designed mechanical product may not be evaluated entirely before being introduced into the market. Thus, any defects may only show themselves as performance issues once the product is in the marketplace. Reliability quantitative (RQ) specifications using proper methodology should be included and evaluated in the product design that meets its expected life before it will be released [1–4].

Mechanical systems such as automobiles, airplanes, and refrigerators [5] convert forms of energy to achieve a specific function (movement of the automobile or airplane and cooling for the refrigerator). The energy conversion requires forces and movement of components, which eventually produce the desired functions with multiple system mechanisms. In the process, mechanical systems are typically subjected to repeated loads. Most mechanical products are made of multi-module structures. If the modules are properly designed and assembled, mechanical systems can work properly and perform their planned functions. For example, in utilizing the vapor-compression refrigeration cycle, a domestic refrigerator is used to cool or freeze food. The refrigerator evaporator provides cooled air to both the refrigerator and freezer sections. A refrigerator has multiple subsystems—Door, cabinet, drawers and shelves, control system, compressor, motor, water supplying device, heat exchangers, and other various components. The total number of parts might be as high as 2000. The product lifetime is targeted to have no less than a B20 life 10 years. As a refrigerator consists of 8 to 10 modules (see Figure 1a) and each module may contain as many as 100 components, the lifetime target of each module needs to have a B1 life of 10 years. The product lifetime of the system is determined by the module with the shortest life, which is module #3 in Figure 1b.

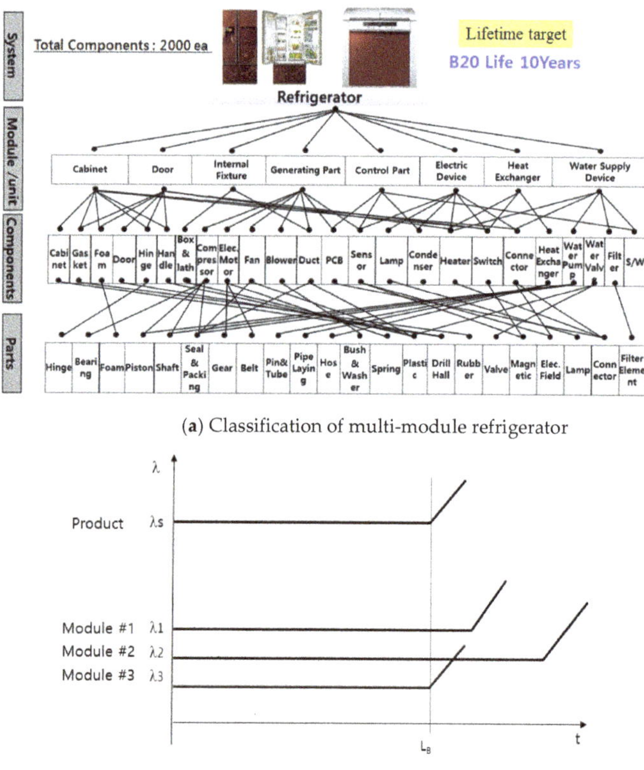

(a) Classification of multi-module refrigerator

(b) Product lifetime L_B and failure rate λ_s

Figure 1. Product lifetime with multi-modules decided by newly designed module (**a**) Classification of multi-module refrigerator; (**b**) Product lifetime L_B and failure rate λ_s.

To avoid the failure of a mechanical system in the field [6,7], it should be designed to robustly endure or survive whatever usage conditions the customers subject the system. Design faults should be recognized and altered by statistical methodology [8] or reliability testing [9] before a commercially manufactured goods is released. However, the statistical methodology and subsequent testing may require large numbers of computations for an optimum solution but may not identify the ultimate failures that may occur in the field by consumers. If there are design faults that cause an inadequacy of strength (or stiffness) when a system is subjected to repeat loading, the system will fail before its expected lifetime due to fatigue failure. American Society for Testing and Materials (ASTM) procedures typically require large samples, skilled personnel, testing apparatus arranged by data acquisition systems, etc. Thus, discovering possible mechanical failures such as fatigue can involve time-consuming and costly testing procedures [10–13]. To recognize these limitations, there have been numerous attempts to systematically evolve fatigue testing [14]. It is difficult to estimate the lifetime cycles of problematic parts in multi-module products where failures rarely occur in the field due to design flaws.

An alternative method, based on reliability block diagrams [15], is where the accelerated life testing (ALT) could be scrutinized [16–24]. It includes test planning for the product, failure mechanics, accelerated procedures, sample size equation, etc. Elsayed [25] categorized physics/statistics, statistical, and physics/experimental-based models for assessment. Meeker [26] proposed numerous feasible recommendations to organize an ALT. Carrying out an ALT [27,28] requires numerous concepts such as the BX life for the product test plan based on reliability engineering, a life-stress model, sample size equation, and fracture mechanics [29–32] because failure may occur suddenly from the frail components in a system. Contemporary experimental methods may fail to reproduce the design defects. These methods may evaluate insufficient part samples in multi-module products and may not identify the failure(s) that actually happen in the marketplace.

To implement the optimal design of a mechanical system, engineers have relied on traditional design approaches such as strength of materials [33]. A recent fracture mechanics study proposed that the crucial elements might be fracture toughness as an alternative of strength as an applicable material property. As quantum mechanics has been used in electronic technology, designers have identified system failures from micro-void coalescence (MVC) and noted a great number of metallic alloys or numerous engineering plastics [34]. To determine the failure phenomena of a mechanical system, a better life-stress model might be combined with the traditional design approaches and applicable methodology of identifying the failure of electronic parts due to small cracks or pre-existing defects. This approach would not be feasible to model using current finite element methods (FEMs) [35].

To better identify product failures in the marketplace, there is another engineering perspective that incorporates the FEM [36]. Many engineers believe that rare system failures might be evaluated by: (1) Mathematical modeling utilizing Newtonian or Lagrangian techniques; (2) obtaining the system stress/strain from the time response for (dynamic) loads; (3) making use of the rain-flow counting method for von-Mises stress [37,38]; and (4) approximating system damage by Palmgren–Miner's rule [39]. Nevertheless, utilizing a systematic method that can give closed-form, precise solutions would involve utilizing numerous assumptions that might not identify multi-module system failures, due to material defects such as micro-voids and contacts when subjected to loads.

This study introduces a parametric ALT as a systematic reliability method that can generate the RQ specifications such as mission time for identifying and modifying the design faults of newly designed mechanical systems. It incorporates: (1) An ALT plan formed on system BX lifetime, (2) a load examination for ALT, (3) customized ALTs with the alterations, and (4) an assessment of whether the last design(s) of the system fulfills the objective BX lifetime. A newly designed hinge kit system (HKS) in a domestic refrigerator subjected to repeated impact loading is provided as an example.

2. Parametric ALT for Mechanical System

2.1. Definition of BX lifetime for Putting a Whole Parametric ALT Plan

To carry out a parametric ALT, the BX life as a measure of system lifetime is required. The BX life, L_B, can be explained as the elapsed time at which X percent of a collection of a selected product might have failed. Otherwise, 'BX life Y years' is a good expression for product lifetime that helps to satisfactorily decide the cumulative failure rate of a product and respond to field circumstances. For instance, if the lifetime of a product has a B20 life of 10 years, then 20% of the population might have been unsuccessful in achieving one's goal for 10 years of the working period.

Reliability might be explained as the system's ability to work under specified conditions for a stated period of time [40]. Product reliability, as shown in Figure 2, is often illustrated with the "bathtub curve" that is composed of three sections [41]. First, there is a declining failure rate in the earlier product life ($\beta < 1$). Secondly, there is a constant failure rate ($\beta = 1$) in the middle of the product's life. Lastly, there is a growing failure rate at the end of the product life ($\beta > 1$). If a manufacturer produces a product whose failure rate follows the bathtub curve, it might have difficulties achieving success in the marketplace because of shorted lifetime and large failure rates due to design faults in the early product life. Manufacturers need to enhance the product design by increasing its reliability targets to (1) eliminate untimely failures, (2) lessen random failures over the product lifetime, and (3) lengthen system lifetime. As the design of a mechanical product improves, its failure rate in the marketplace should decease and the product lifetime should be extended. For such circumstances, the conventional bathtub curve might be changed to a straight line in Figure 2.

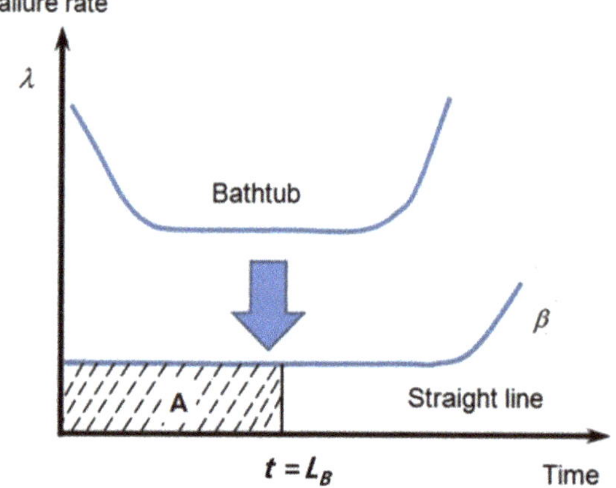

Figure 2. Bathtub curve and straight line.

The failure rate on the bathtub (or straight line) can be defined as

$$\lambda = \frac{f}{R} = \frac{dF/dt}{R} = \frac{(1-R)'}{R} = \frac{-R'}{R} \tag{1}$$

where λ is the failure rate, f is the failure density function, R is reliability, and F is unreliability.

If Equation (1) is integrated over time, we can obtain the $X\%$ cumulative failure $F(L_B)$ at BX life, L_B. That is,

$$\int \lambda dt = -\ln R \tag{2}$$

That is to say, it can be expressed as:

$$A = \langle \lambda \rangle \cdot L_B = \int_0^{L_B} \lambda(t) \cdot dt = -\ln R(L_B) = -\ln(1 - F) \cong F(L_B) \tag{3}$$

where L_B is the BX life, A is the area that can be obtained from the multiplication of failure rate, λ, and BX life, L_B.

Consequently, if a product failure follows an exponential distribution, the reliability of a mechanical product can be defined as:

$$R(L_B) = 1 - F(L_B) = e^{-\lambda L_B} \cong 1 - \lambda L_B \tag{4}$$

Equation (4) is relevant for when there are less than approximately 20% of the cumulative failures for the system [42]. The mechanical system could be improved by obtaining the objective product lifetime, L_B, and failure rate, λ, after optimally identifying the market failure by parametric ALT and modifying the problematic design (or material) of structures (Figure 3).

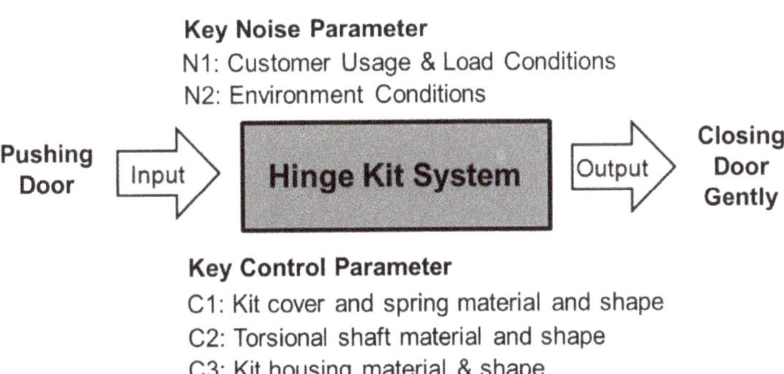

Figure 3. Parameter diagram of hinge kit system (HKS) (example).

In seeking to improve the lifetime target of a mechanical system through an ALT examination, there are three potential product modules: (1) An altered module, (2) a newly designed module, and (3) an alike module to the previous design base on demand in the marketplace. The newly designed HKS in the refrigerator examined here as a case study was a new module that had design faults that had to be rectified because customers asked for replacements with a new one because the product failed during its expected lifetime.

The new module *B* from the market data shown in Table 1 had a failure rate of 0.24% per year and a B1 life of 4.2 years. To answer customer requests, a new lifetime target for the HKS was set to have B1 life 10 years with a cumulative failure rate of one percent.

2.2. Failure Mechanics and Accelerated Testing for Design

Mechanical systems typically move energy and power from one location to another through mechanical mechanisms. If there is a design fault in the structure that causes an inadequate strength (or stiffness) when the loads are exerted, the mechanical system may suddenly fail before its anticipated lifetime. Fatigue due to design flaws can be characterized by two factors: (1) the stress due to loads on the structure and (2) the type of materials (or shape) used in the product. In reproducing the system failure by a parametric ALT, a designer could optimally design components with proper shapes and materials. The product could sustain repetitive loads over its lifetime so that it could achieve the targeted reliability (Figure 4).

Table 1. Whole ALT plan of mechanical system such as modules in a refrigerator.

Modules	Market Data		Expected Reliability				Targeted Reliability	
	Yearly Failure Rate, %/Year	BX Life, Year	Yearly Failure Rate, %/Year			BX Life, Year	Yearly Failure Rate, %/Year	BX Life, Year
A	0.35	2.9	Similar	×1	0.35	2.9	0.10	10(BX = 1.0)
B	0.24	4.2	New	×5	1.20	0.83	0.10	10(BX = 1.0)
C	0.30	3.3	Similar	×1	0.30	3.33	0.10	10(BX = 1.0)
D	0.31	3.2	Modified	×2	0.62	1.61	0.10	10(BX = 1.0)
E	0.15	6.7	Modified	×2	0.30	3.33	0.10	10(BX = 1.0)
Others	0.50	10.0	Similar	×1	0.50	10.0	0.50	10(BX = 5.0)
Product	1.9	2.9	-	-	3.27	0.83	1.00	10(BX = 10)

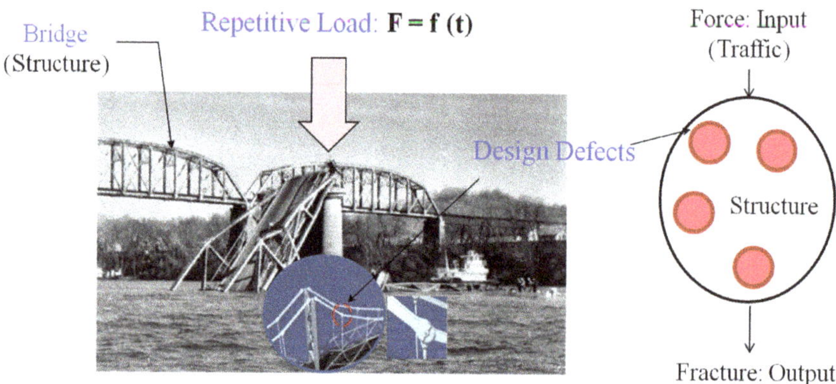

Figure 4. Fatigue failure on the product produced by (random) repeated load and design flaws.

The most important issue for a reliability test is how quickly the possible failure mode might be obtained. A failure model must be derived and its associated coefficients determined. The life-stress (LS) model also incorporates stresses and reaction parameters. The generalized life-stress (LS) model [1,43,44] might thus be defined as

$$TF = A[\sinh(aS)]^{-1} \exp\left(\frac{E_a}{kT}\right) \tag{5}$$

The sine hyperbolic expression $[\sinh(aS)]^{-1}$ in Equation (5) can be expressed as:

1. $(S)^{-1}$ in Equation (5) has a little linear effect at first,
2. $(S)^{-n}$ in Equation (5) has what is regarded as a medium effect, and
3. $(e^{aS})^{-1}$ in Equation (5) is big in the end.

An ALT is normally performed in the medium range, and Equation (5) might be defined as

$$TF = A(S)^{-n} \exp\left(\frac{E_a}{kT}\right) \tag{6}$$

As the stress of a mechanical system may not be easy to measure during testing, Equation (6) must be redefined. When the power is defined as the multiplication of flows and effort, stresses may come from effort in a multi-port system (Table 2) [45].

Table 2. Power definition in a multi-port system.

System	Effort, *e(t)*	Flow, *f(t)*
Mechanical translation	Force, *F(t)*	Velocity, *V(t)*
Mechanical rotation	Torque, *τ(t)*	Angular velocity, *ω(t)*
Compressor, Pump	Pressure difference, $\Delta P(t)$	Volume flow rate, *Q(t)*
Electric	Voltage, *V(t)*	Current, *i(t)*
Magnetic	Magneto-motive force, e_m	Magnetic flux, φ

Stress is a physical quantity that indicates the internal forces that adjacent particles of a continual material apply on each other. For a mechanical system, because stress comes from effort, Equation (6) might be redefined as

$$TF = A(S)^{-n} \exp\left(\frac{E_a}{kT}\right) = B(e)^{-\lambda} \exp\left(\frac{E_a}{kT}\right) \tag{7}$$

where A and B are constants

To derive the acceleration factor (AF) that can mainly enfluence the assessment of fatigue strength in product, expressed as the inverse of the stress ratio, R (=$\sigma_{min}/\sigma_{max}$), from Equation (7), AF might be expressed as the proportion between the adequate elevated stress amounts and normal working conditions. AF might be altered to incorporate the effort ideas:

$$AF = \left(\frac{S_1}{S_0}\right)^n \left[\frac{E_a}{k}\left(\frac{1}{T_0} - \frac{1}{T_1}\right)\right] = \left(\frac{e_1}{e_0}\right)^\lambda \left[\frac{E_a}{k}\left(\frac{1}{T_0} - \frac{1}{T_1}\right)\right] \tag{8}$$

2.3. Parametric ALT of Mechanical Systems

To obtain the mission cycle of ALTs from the objective BX lifetime on the experiment scheme in Table 1, the sample size formulation integrated with the AF should be obtained [1]. Until now, numerous methodologies have been suggested to decide sample size. The Weibayes model for Weibull analysis is a popularly recognized method of examining reliability data. However, it is hard to directly use because of the mathematical complication. The whole cases as failures ($r \geq 1$) and no failures ($r = 0$) need to be separated. Consequently, it is possible to acquire a comprehensible sample size equation that might provide the mission cycle after proper assumptions.

In choosing the model parameters to maximize the likelihood function, the maximum likelihood estimation (MLE) statistic is a widespread way of approximating the parameters of a model. The characteristic life η_{MLE} can be expressed as:

$$\eta_{MLE}^{\beta} = \sum_{i=1}^{n} \frac{t_i^{\beta}}{r} \tag{9}$$

where η_{MLE} is the maximum likelihood estimate of the characteristic life, *n* is the total number of samples, t_i is the test duration for each sample, and *r* is the number of failures.

If the number of failures is $r \geq 1$ and the confidence level is $100(1 - \alpha)$, the characteristic life, η_α, can be approximated from Equation (9),

$$\eta_\alpha^{\beta} = \frac{2r}{\chi_\alpha^2(2r+2)} \times \eta_{MLE}^{\beta} = \frac{2}{\chi_\alpha^2(2r+2)} \times \sum_{i=1}^{n} t_i^{\beta} \, \text{for} \, r \geq 1 \tag{10}$$

where $\chi_\alpha^2()$ is the chi-square distribution when the *p*-value is α.

Assuming there are no number of failures, $\ln(1/\alpha)$ is mathematically identical to the chi-square value, $\frac{\chi_\alpha^2(2)}{2}$ [46]. In other words,

$$p-value: \alpha = \int_{\chi_\alpha^2(2)}^{\infty} \left(\frac{e^{-\frac{x}{2}} x^{\frac{\nu}{2}-1}}{2^{\frac{\nu}{2}} \Gamma\left(\frac{\nu}{2}\right)} \right) dx = \int_{2\ln\alpha^{-1}}^{\infty} \left(\frac{e^{-\frac{x}{2}} x^{\frac{\nu}{2}-1}}{2^{\frac{\nu}{2}} \Gamma\left(\frac{\nu}{2}\right)} \right) dx \, for x \geq 0 \tag{11}$$

where Γ is the gamma function and ν is the shape parameter

For $r = 0$, the characteristic life η_α from Equation (10) can be defined as:

$$\eta_\alpha^\beta = \frac{2}{\chi_\alpha^2(2)} \times \sum_{i=1}^{n} t_i^\beta = \frac{1}{\ln\frac{1}{\alpha}} \times \sum_{i=1}^{n} t_i^\beta \tag{12}$$

As Equation (10) is proved for all cases $r \geq 0$, characteristic life, η_α, can be expressed as follows:

$$\eta_\alpha^\beta = \frac{2}{\chi_\alpha^2(2r+2)} \times \sum_{i=1}^{n} t_i^\beta \, for r \geq 0 \tag{13}$$

If the logarithm in the Weilbull distribution is taken, the connection between characteristic life and BX life, L_B, can be defined as:

$$L_B^\beta = \left(\ln \frac{1}{1-x} \right) \times \eta^\beta \tag{14}$$

If the approximated characteristic life of the p-value α, η_α, in Equation (13), is changed into Equation (17), we obtain the BX life formulation:

$$L_B^\beta = \left(\ln \frac{1}{1-x} \right) \times \frac{2}{\chi_\alpha^2(2r+2)} \times \sum_{i=1}^{n} t_i^\beta \tag{15}$$

As nearly all life testing commonly has inadequate samples to approximate the lifetime for the assigned number of failures that might be less than that of the sample size, the plan testing time can begin as:

$$nh^\beta \geq \sum t_i^\beta \geq (n-r) \times h^\beta \tag{16}$$

If Equation (16) is exchanged with Equation (15), the BX life equation can be redefined as:

$$L_B^\beta \cong \left(\ln \frac{1}{1-x} \right) \times \frac{2}{\chi_\alpha^2(2r+2)} \cdot nh^\beta \geq \left(\ln \frac{1}{1-x} \right) \times \frac{2}{\chi_\alpha^2(2r+2)} \times (n-r)h^\beta \geq L_B^{*\beta} \tag{17}$$

If Equation (17) is rearranged, the sample size formulation with the failure numbers can be defined as:

$$n \geq \frac{x\chi_\alpha^2(2r+2)}{2} \times \frac{1}{\left(\ln \frac{1}{1-x} \right)} \times \left(\frac{L_B^*}{h} \right)^\beta + r \tag{18}$$

Because $\frac{\chi_\alpha^2(2r+2)}{2} \cong (r+1)$ for $\alpha = 0.6$ and $\ln(1-x)^{-1} = x + \frac{x^2}{2} + \frac{x^3}{3} + \cdots \cong x$, the sample size Equation (21) can be simply close to:

$$n \geq (r+1) \times \frac{1}{x} \times \left(\frac{L_B^*}{h} \right)^\beta + r \tag{19}$$

where the sample size equation can be restated as $n \sim$ (failure numbers + 1)·(1/cumulative failure rate)·((target lifetime/(plan testing time)) ^ β + r.

If Equation (8) is attached to the plan testing time h, Equation (19) can replaced as:

$$n \geq (r+1) \times \frac{1}{x} \times \left(\frac{L_B^*}{AF \cdot h_a} \right)^\beta + r \tag{20}$$

If the lifetime target of a mechanical system such as the HKS in a domestic refrigerator is assigned to be B1 life 10 years, the mission cycles might be attained for an assigned set of samples subjected to the food loading. In ALTs, the design flaws of the new product might be recognized to fulfill the lifetime target [47–49].

2.4. Case Study—Reliability Design of a Newly Designed HKS in Domestic Refrigerator

When a consumer operates a refrigerator door, they want to comfortably close the door. A new HKS was designed for the refrigerator (see Figure 5) to enhance the ease of opening and closing the door for the consumer. When opening/closing the door, the HKS was subjected to repeated impact loads over the lifetime of the domestic refrigerator. To endure the loads of the HKS, new metals—standard austenitic ductile iron (18 wt% Ni)—for the torsional shaft were a key metal component [50] used. Due to their cheap cost as well as outstanding workability, ductile cast irons have been utilized for numerous mechanical parts. They have fine monotonic strength and high ductility compared to malleable cast irons and gray cast irons. The fatigue strength of ductile cast irons is comparatively lower than those of the steels and alloys with the identical quantity of monotonic strength because of their distinctive microstructure holding graphite particles and casting defects [51]. The fatigue strength of a ductile cast iron in the current HKS design was evaluated through parametric ALT.

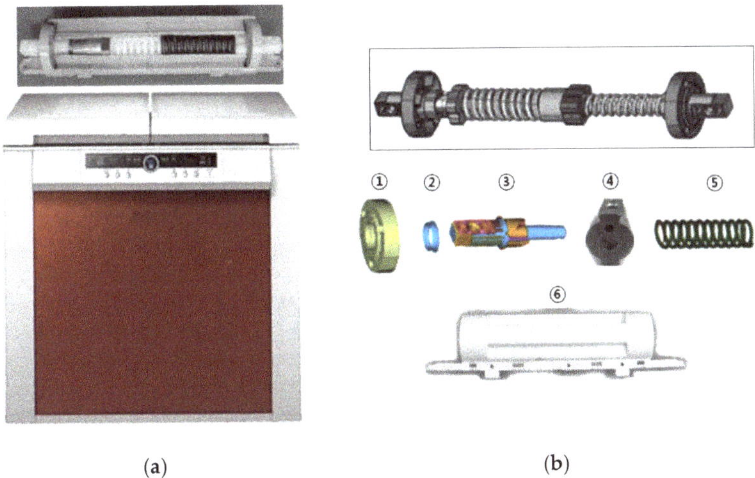

<table>
<tr><td align="center">(a)</td><td align="center">(b)</td></tr>
</table>

Figure 5. A domestic refrigerator (**a**) and HKS (**b**) and its parts: (1) kit cover, (3) support, (4) torsional shaft (cast iron), (5) spring, and (6) kit housing (high-impact polystyrene, HIPS).

The HKS shown in Figure 5b consisted of a kit cover, torsional shaft (ductile iron), spring, and kit housing. To suitably work its function for a product lifetime, the HKS should be designed to endure the working circumstances subjected to it by the customers who utilize the refrigerator. In the Korean domestic market, the representative customer opened and closed the refrigerator door from three to ten times per day. Stocking food in the refrigerator had some repeated working procedures: (1) Open the door of refrigerator, (2) put the food into it, and then (3) close it. The HKS had different mechanical impact loadings when the customer utilized it.

The HKS in the marketplace had been fracturing, causing customers to demand the refrigerator be replaced. As subject to repeated impact stresses in using the refrigerator door, it was determined that the problematic HKS originated from several design defects. Market data also indicated that the returned products had crucial design problems on the structure, including stress risers—sharp corner angles and thin ribs. These design defects prohibited the HKS from enduring the repeated impact loads during the openings/closings and resulted in a crack that propagated to its end. The HKS was originally designed to endure repeated impact loading under customer working conditions (Figure 6).

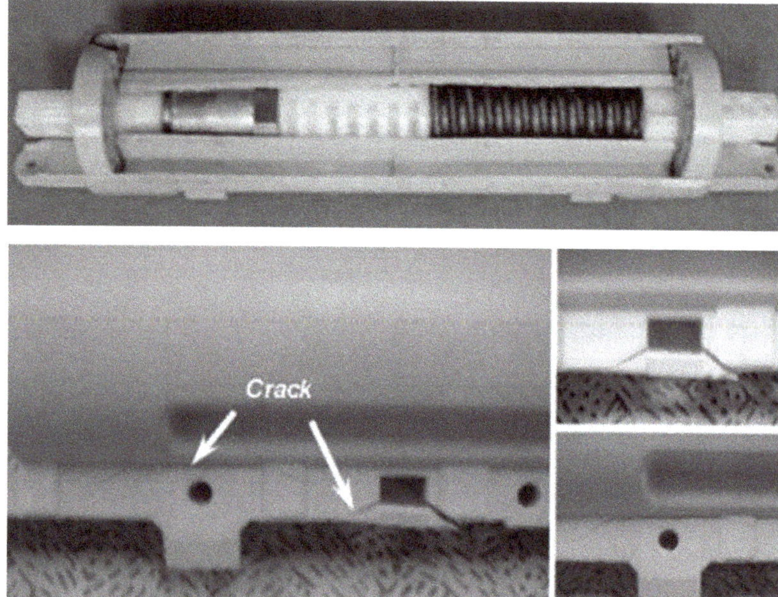

Figure 6. Damaged HKS in field after use.

When customers operated the refrigerator door, they could take out and put in food. Relying on the end-user working conditions, the HKS experienced repeated impact loading in the process. To correctly work the HKS, many mechanical structural parts in the HKS assembly needed to be designed robustly. As the concentrated stress in the mechanical system was revealed at stress raisers such as sharp corner angles, it was crucial to demonstrate these design flaws experimentally. As a result, engineers could then modify the design.

As seen in Figure 7, from the functional design ideas of a mechanical HKS, we knew that the impact force on the HKS came from the door weight. That is, the moment balance around HKS can be stated as

$$M_0 = W_{door} \times R \tag{21}$$

$$(21) = T_0 = F_0 \times R \tag{22}$$

where b is distance from the HKS to the center of gravity (CG) of the door.

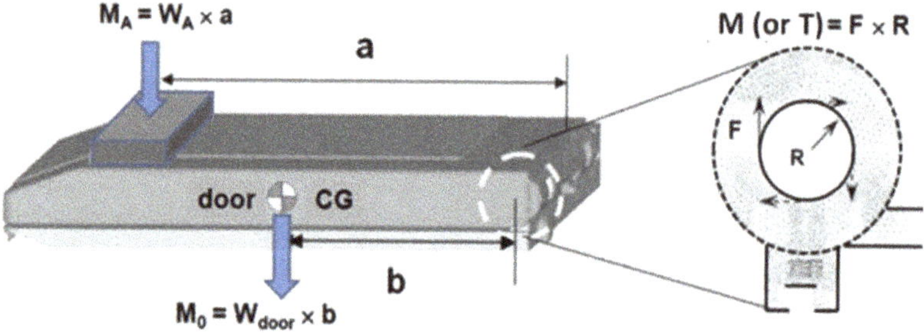

Figure 7. Functional design ideas of a mechanical HKS.

To increase the impact on the HKS, additional accelerated weight was added. The moment balance around the HKS with an accelerated weight can be stated as

$$M_1 = M_0 + M_A = W_{door} \times b + M_A \times a \tag{23}$$

$$(23) = T_1 = F_1 \times R \tag{24}$$

where a is distance from the HKS to the accelerated weight

Because the time to failure depended on the impact force due to moment, the impact was controlled during the accelerated life testing. Under the same working conditions, the life-stress model (LS model) in Equation (7) can be restated as

$$TF = A(S)^{-n} = AT^{-\lambda} = A(F \times R)^{-\lambda} = B(F)^{-\lambda} \tag{25}$$

where A and B are constant

Therefore, the AF in Equation (8) can be restated as

$$AF = \left(\frac{S_1}{S_0}\right)^n = \left(\frac{T_1}{T_0}\right)^\lambda = \left(\frac{F_1}{F_0}\right)^\lambda \tag{26}$$

For a refrigerator including the HKS, the environmental (or working) customer conditions were roughly 0–43 °C with a relative humidity varying from 0 to 95%, and 0.2–0.24 g of acceleration. As previously mentioned, the number of openings/closings of the HKS per day varied from 3 to 10 times. With a design criterion of a product lifetime for 10 years, L_B^*, the HKS has 36,500 usage cycles in the worst case.

Under a lifetime target—B1 life 10 years—if the number of lifetime cycles L_B^* and AF are computed for the assigned sample size, the actual mission cycles, h_a, might be acquired from Equation (20). Then, the ALT equipment can be made and performed in accordance with the working course of the HKS. Through parameter ALTs, the design missing parameters (or design flaws) for the new mechanical system can be identified.

The greatest impact force due to the door weight exerted by the customer in utilizing the refrigerator, F_1, was 1.1 kN. To determine the stress level for ALT, we used the step-stress life test that can assess the lifetime under constant used-condition for various accelerated weights [52]. As the stress level to a different level was changed, the failure times of the HKS at a particular stress level was observed. Finally, for an ALT with an accelerated weight, we determined that the exerted impact force, F_2, was 2.76 kN. With a cumulative damage exponent, λ, of 2, the AF was 6.3 from Equation (26). To obtain the missing design parameters of a newly designed HKS, a lifetime target should be more than B1 life 10 years. If the shape parameter β was 2.0, the number of test cycles computed from Equation (20) would be 23,000 cycles for 6 sample units. If the parametric ALT failed less than once for 23,000 cycles, the lifetime for the HKS would be assured to be B1 life 10 years (Figure 8).

The control console was used to run the testing apparatus—the number of test cycles, beginning or ending the equipment, etc. As the start knob on the controller console gave the starting signal, the straight hand-shaped arms clasped and raised the refrigerator door. When the door was shut, it was exerted to the HKS with the greatest mechanical impact force due to the accelerated load (2.76 kN).

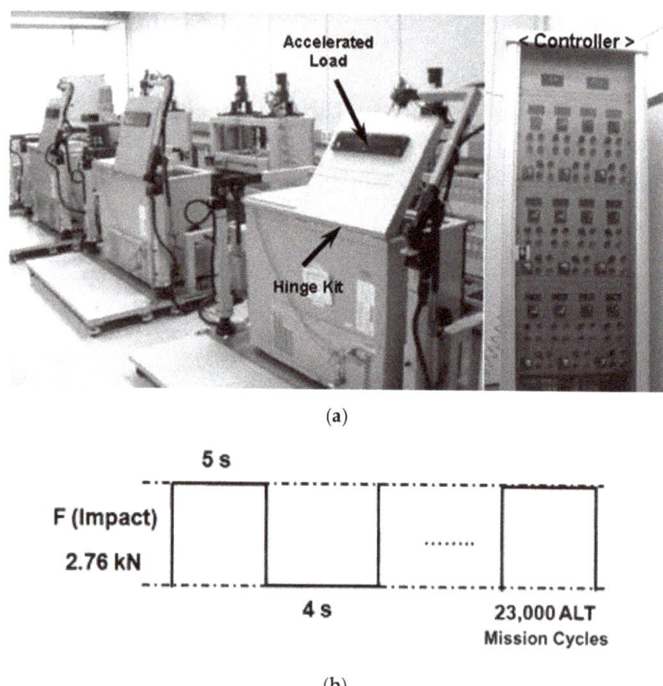

(a)

(b)

Figure 8. ALT (**a**) equipment; (**b**) duty cycles of repeated impact load F.

3. Results and Discussion

In the 1st ALT, the housing of the HKS failed at 3000 cycles. Figure 9 shows the failed product from the marketplace and the 1st ALT. Upon carefully observing the failure locations from the marketplace and the first ALT, it was found that the failures were around the housing and its support in the HKS structure as a consequence of high impact stress.

Figure 10 provides a graphical presentation of the 1st ALT results and the failure data from the field shown on the Weibull plot. As the two patterns had similar slopes on the plot, each loading state of the 1st ALT and the field over the product lifetime were alike under the operational conditions of customers. Thus, it should be expected that the test samples will fail like those in the field. For the shape parameter, β, the final shape parameter from the chart was affirmed to be 2.0, compared with the estimated value—2.0. Based on both test results in the Weibull plot, the parametric ALT was effective because it identified the design flaws that were accountable for the field failures. In other words, as substantiated by two items—the visual representation in the pictures and similar slopes in the Weibull plot—these systematic methods were well-founded in identifying the problematic designs that accounted for the failures from the field. These failures decided the product (refrigerator's) lifetime.

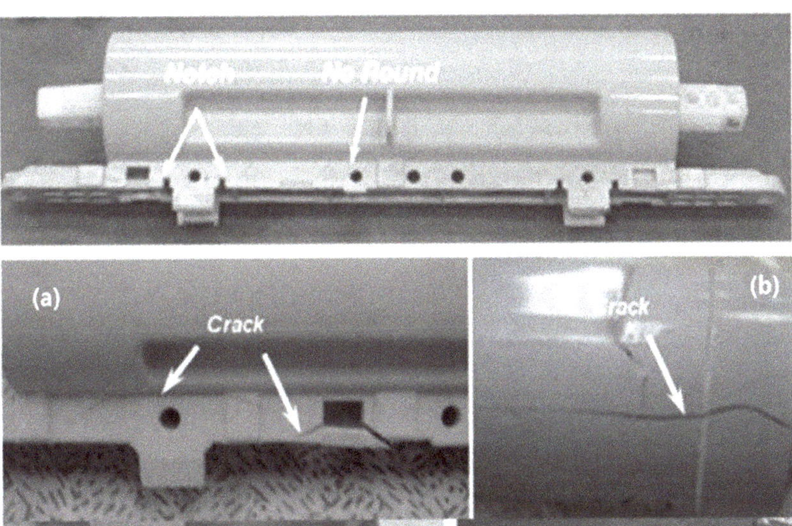

Figure 9. Failed HKS from the marketplace and in the first ALT: (**a**) Failed product after first ALT; (**b**) product with crack in field.

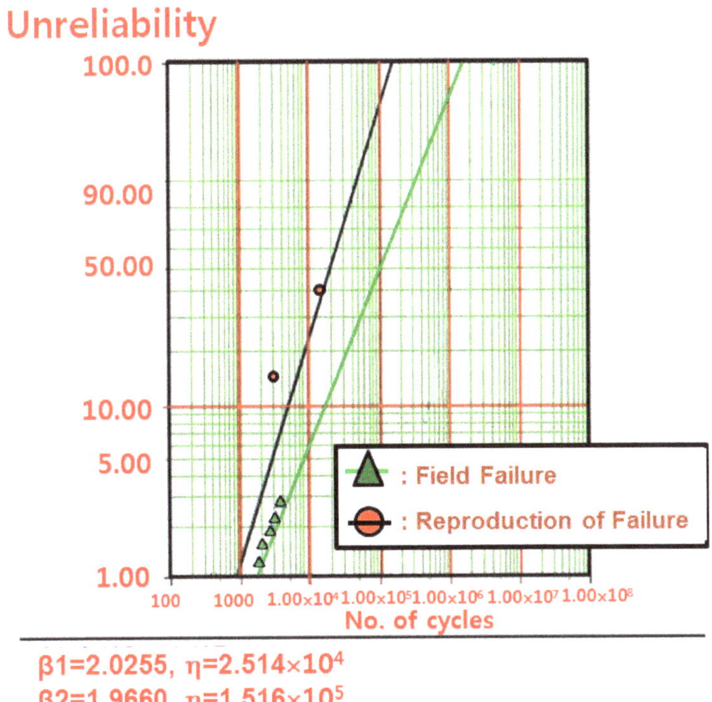

Figure 10. Market data and outcomes of 1st ALT on Weibull plot.

Due to the design defect of no support in the high-stress areas, the repeated impact loading in conjunction with this structural defect may have produced fracturing of the

HKS housing. This design defect can be altered by adding the support ribs, C1 (Figure 11). Stress analysis, which can be combined with fatigue analysis and parametric ALT, was carried out by using a finite element analysis (FEA). When the HKS was fixed against the wall (or surface) as the boundary conditions, the straightforward impact loads, as seen in Figure 7, were applied. Using materials and processing conditions similar to those of the finished HKS, the constitutive properties of the materials such as HIPS (HKS housing) were determined. The maximum stresses for the old and new designs were estimated separately. Based on these results, the appropriateness of the current designs for the HKS housing was evaluated. After modifying the new designs to improve the design against fatigue, the estimated stress concentrations in the HKS housing decreased from 21.2 to 15.0 MPa using the FEM analysis. It was expected that this new design should be effective in reducing fatigue failure of the HKS housing when subjected to repeated load under the consumer usage conditions.

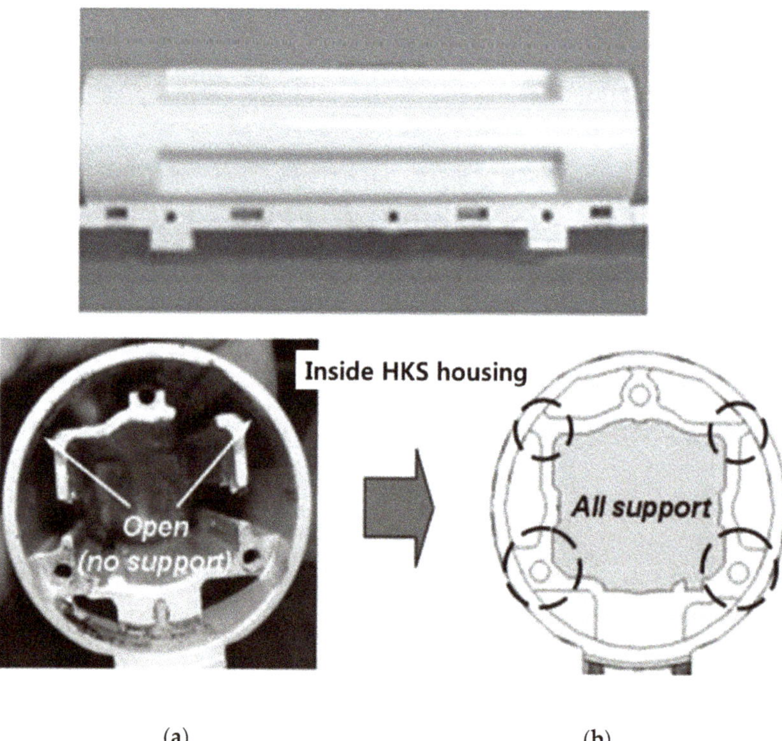

(a)   (b)

Figure 11. Failed HKS housing in the first ALT: (**a**) Its root cause; (**b**) design modifications.

With the confirmed shape parameter β of 2.0, the real mission cycles computed from Equation (20) were 23,000 cycles for the six sample units. If the HKS failed at less than once for 23,000 cycles, its lifetime would be assured to be B1 life 10 years. As seen in Figure 12, in the second ALT, from the outside corner, the torsional shaft in the HKS that was made of ductile iron failed at 12,000 cycles. Such ductile cast iron accounts for a major family of metals that are extensively used for gears, automobile crankshafts, dies, and numerous machine parts because of its good machinability, fatigue strength, and high modulus of elasticity. They have a mass fraction (%) as follows: Carbon (3.0–3.7), silicon (1.2–2.3), manganese (0.25), magnesium (0.07), phosphorus (0.03) [53].

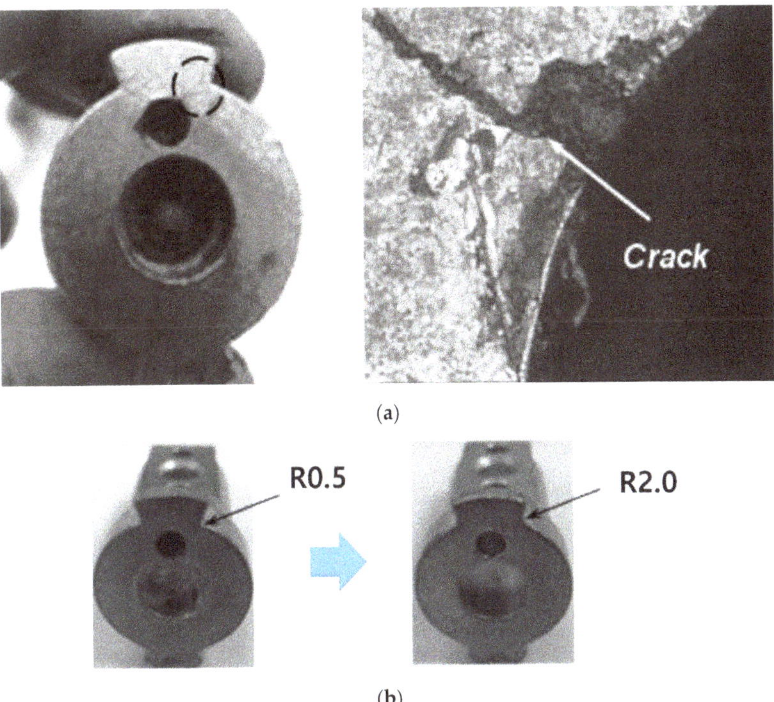

(a)

(b)

Figure 12. Root cause and design modification of cracked torsional shaft (cast iron) in the second ALT: (**a**) Root cause; (**b**) design modification.

When closely examining the product failure in the 2nd ALT, the torsional shaft in the HKS had insufficient strength to endure the repeated impact loading of the opening/closing of the door. When subjected to repeated impact loads, the stress amplification of mechanical components such as the torsional shaft in the HKS not only occurred at minute defects or cracks on a microscopic level of material but could also happen in stress concentrations such as in sharp corners, fillets, holes, and notches on the macroscopic range that are normally explained as stress raisers. For example, the stress concentration at the sharp-edged corners depended on fillet radius [54]. To improve the design, the torsional shaft was altered by giving it more rounding from R0.5 mm to R2.0 mm, C2 (Figure 12).

For the HKS upgrade, the design basis of new samples was determined to be more than the lifetime target—B1 life 10 years. To confirm the design of the HKS, a 3rd ALT was performed. As the affirmed value, β, on the Weibull plot was 2.0, for the lifetime target—B1 life 10 years—the actual mission cycles in Equation (20) were 23,000 for the six-sample size. In the third ALT, there were no design issues in the HKS until the experiment reached 23,000 cycles. It was therefore concluded that the altered design parameters obtained from the 1st and 2nd ALTs were successful.

Table 3 provides a summary of the ALT results. With the alternative designs, the HKS was assured to have a lifetime target—B1 life 10 years. That is, we knew that the product would have 99% reliability (or 1% unreliability) for 10 years with a yearly failure rate of 0.1%.

Table 3. Results of ALTs.

Parametric ALT	1st ALT	2nd ALT	3rd ALT
	Initial Design	Second Design	Last Design
In 23,000 cycles, there are no problems in the HKS	3000 cycles: 2/6 Fracture (HKS Housing)	12,000 cycles: 4/6 crack (Torsional Shaft)	23,000 cycles:6/6 OK 41,000 cycles:6/6 OK
HKS Structure	Fracture	Crack	-
Action plans	C1: No → 2 support ribs	C2: R0.5mm → R2.0mm Roundness corner of torsional shaft	-

4. Conclusions

A systematic reliability method was proposed for a new mechanical system such as an HKS in refrigerators. It incorporated: (1) A parametric ALT plan formed on product BX lifetime, (2) a load examination for ALT, (3) customized ALTs with the design alterations, and (4) an assessment of whether the last design(s) of the product fulfilled the objective BX lifetime. Testing was conducted to subject the HKS in the domestic refrigerator to repeated impact loading.

In the first ALT, the HKS housing had insufficient strength for repeated impact loading and cracked. As an action plan, these flaws in the HKS were corrected by adding supporting ribs. In the second ALT, the torsional shaft made of ductile iron cracked. Due to its good machinability, fatigue strength, and high modulus of elasticity, iron is widely used in machine components. To improve its strength for impact loading, the torsional shaft was altered by increasing the corner roundness.

With these altered design parameters, in the third ALT, there were no design issues. The altered design parameters were assured to satisfy the lifetime need of the HKS—B1 life 10 years. With the examination of returned products from the marketplace, laboratory load evaluations and testing, and parametric ALTs with design modifications, the design flaws were identified and remedied to create a robust design with a remarkably lengthy lifetime. This parametric ALT is also recommended to be applied to other metals that can be used in the numerous machine parts such as cam, gears, automobile crankshafts, and dies [55–59].

Author Contributions: S.W. conducted the conceptualization, methodology, analysis, and testing and wrote the article. D.L.O. checked the analysis and writing for original draft. M.M.T. and S.M.A. review and edit the paper. D.E.W. supervised the research and paper preparation. All authors have read and agreed to the published version of the manuscript.

Funding: This research received no external funding.

Institutional Review Board Statement: Not applicable.

Informed Consent Statement: Not applicable.

Data Availability Statement: The data presented in this study are available on request from the corresponding author.

Conflicts of Interest: The authors state no conflicts of interest.

Abbreviations

BX	time that is an accumulated failure rate of $X\%$: durability index
E_a	activation energy, eV
e	effort
f	flow
F	impact force, kN
$F(t)$	unreliability
h	testing cycles (or cycles)
$h*J$	nondimensional testing cycles, $h^* = h/L_B \geq 1$junction equation
k	Boltzmann's constant, 8.62×10^{-5} eV/deg
L_BM	target BX life and $x = 0.01X$, on the condition that $x \leq 0.2$ moment around the hinge kit system, kN$\times$ m
N	number of test samples
Q	amount of energy absorbed or released during the reaction. For the semiconductor total number of dopants per unit area
RRr	radius of the hinge kit system, mratio for minmum stress to maximum stress in stress cycle, $\sigma_{min}/\sigma_{max}$failed numbers
S	stress
TTt_i	torque around the hinge kit system, kN$\cdot$ mtemperature, Ktest time for each sample
TF	time to failure
X	accumulated failure rate, %
xW_AW_{door}	$x = 0.01X$, on condition that $x \leq 0.2$.accelerator weight, kgdoor weight, kg
Greek symbols	
ζ	electrical field applied
η	characteristic life
$\lambda\chi^2\alpha$	cumulative damage exponent in Palmgren–Miner's rulechi-square distribution confidence level
Superscripts	
β	shape parameter in Weibull distribution
n	stress dependence, $n = -\left[\frac{\partial \ln(T_f)}{\partial \ln(S)}\right]_T$
Subscripts	
0	normal stress conditions
1	accelerated stress conditions

References

1. Woo, S.; Pecht, M.; O'Neal, D. Reliability design and case study of the domestic compressor subjected to repetitive internal stresses. *Reliab. Eng. Syst. Saf.* **2020**, *193*, 106604. [CrossRef]
2. Deming, W.E. *Elementary Principles of the Statistical Control of Quality*; Japan JUSE: Hyatt Regency, Tokyo, 1950.
3. CMMI Product Team. *Capability Maturity Model Integration (CMMI) Version 2,0, Continuous Representation*; Report CMU/SEI-2002-TR-011; Software Engineering Institute: Pittsburgh, PA, USA, 2018.
4. NASA. *System Engineering Handbook*; NASA/SP-2020-6105 Rev 2; NASA Headquarters: Washington, DC, USA, 2020; p. 92.
5. Cengel, Y.A.; Boles, M.A. *Thermodynamics: An Engineering Approach*, 9th ed.; McGraw-Hill: New York, NY, USA, 2018.
6. Magaziner, I.C.; Patinkin, M. Cold competition: GE wages the refrigerator war. *Harv. Bus. Rev.* **1989**, *89*, 114–124.
7. WRDA 2020 Updates. The Final Report of the US House Committee on Transportation and Infrastructure on the Boeing 737 Max. Available online: https://transportation.house.gov/committee-activity/boeing-737-max-investigation (accessed on 16 September 2020).
8. Chowdhury, S.; Taguchi, S. *Robust Optimization: World's Best Practices for Developing Winning Vehicles*, 1st ed.; John Wiley & Sons Inc.: Hoboken, NJ, USA, 2016.
9. Montgomery, D. *Design and Analysis of Experiments*, 10th ed.; John Wiley: Hoboken, NJ, USA, 2020.
10. ASTM E606/E606M. *Standard Test Method for Strain-Controlled Fatigue Testing*; ASTM International: West Conshohocken, PA, USA, 2019.
11. ASTM E399. *Standard Test Method for Linear-Elastic Plane-Strain Fracture Toughness of Metallic Materials*; ASTM International: West Conshohocken, PA, USA, 2020.
12. ASTM E647. *Standard Test Method for Measurement of Fatigue Crack Growth Rates*; ASTM International: West Conshohocken, PA, USA, 2015.
13. ASTM E739-10. *Standard Practice for Statistical Analysis of Linear or Linearized Stress-Life (S-N) and Strain-Life (ε-N) Fatigue Data*; ASTM International: West Conshohocken, PA, USA, 2015.

14. Braco, R.; Prates, P.; Costa, J.D.M.; Berto, F. New methodology of fatigue life evaluation for multiaxially loaded notched components based on two uniaxial strain-controlled tests. *Int. J. Fatigue* **2018**, *111*, 308–320. [CrossRef]

15. Modarres, M.; Kaminskiy, M.; Krivtsov, V. *Reliability Engineering and Risk Analysis: A Practical Guide*, 3rd ed.; CRC Press: Boca Raton, FL, USA, 2016.

16. Wang, L.; Li, Y. Boundary for aviation bearing accelerated life test based on quasi-dynamic analysis. *Tribol. Int.* **2017**, *116*, 414–421. [CrossRef]

17. Zaharia, S.M. The methodology of fatigue lifetime prediction and validation based on accelerated reliability testing of the rotor pitch links. *Eksploat. Niezawodn.* **2019**, *2*, 638–644. [CrossRef]

18. Chang, M.S.; Lee, C.S.; Choi, B.O.; Kang, B.S. Study on validation for accelerated life tests of pneumatic cylinders based on the test results of normal use conditions. *J. Mech. Sci. Technol.* **2017**, *31*, 2739–2745. [CrossRef]

19. Koo, H.J.; Kim, Y.K. Reliability assessment of seat belt webbings through accelerated life testing. *Polym. Test.* **2005**, *24*, 309–315. [CrossRef]

20. Özsoy, S.; Çelik, M.; Kadıoğlu, F.S. An accelerated life test approach for aerospace structural components. *Eng. Fail. Anal.* **2008**, *15*, 946–957. [CrossRef]

21. Nakada, M.; Miyano, Y. Accelerated testing for long-term fatigue strength of various FRP laminates for marine use. *Compos. Sci. Technol.* **2009**, *69*, 805–813. [CrossRef]

22. Miyano, Y.; Nakada, M.; Ichimura, J.; Hayakawa, E. Accelerated testing for long-term strength of innovative CFRP laminates for marine use. *Compos. Part B Eng.* **2008**, *39*, 5–12. [CrossRef]

23. Rajaneesh, A.; Satrio, W.; Chai, G.B.; Sridhar, I. Long-term life prediction of woven CFRP laminates under three point flexural fatigue. *Compos. Part B Eng.* **2016**, *91*, 539–547. [CrossRef]

24. Bank, L.C.; Gentry, T.R.; Barkatt, A. Accelerated Test Methods to Determine the Long-Term Behavior of FRP Composite Structures: Environmental Effects. *J. Reinf. Plast. Compos.* **1995**, *14*, 559–587. [CrossRef]

25. Elsayed, E.A. *Reliability Engineering*; Addison Wesley Longman: Boston, MA, USA, 1996.

26. Hahn, G.J.; Meeker, W.Q. *How to Plan an Accelerated Life Test (E-Book)*; ASQ Quality Press: Milwaukee, WI, USA, 2004.

27. McPherson, J. Accelerated testing. In *Electronic Materials Handbook Volume 1: Packaging*; ASM International Publishing: Materials Park, OH, USA, 1989; pp. 887–894.

28. McPherson, J. *Reliability Physics and Engineering: Time-to-Failure Modeling*; Springer: New York, NY, USA, 2010.

29. Griffith, A.A. The phenomena of rupture and flow in solids. *Philos. Trans. R Soc. Lond. A* **1921**, *221*, 163–198.

30. Irwin, G. Analysis of stresses and strains near the end of a crack traversing a plate. *J. Appl. Mech.* **1957**, *24*, 361–364.

31. Anderson, T.L. *Fracture Mechanics—Fundamentals and Applications*, 3rd ed.; CRC: Boca Raton, FL, USA, 2017.

32. Neuber, H. Theory of stress concentration for shear strained prismatical bodies with arbitrary non-linear stress-strain law. *J. Appl. Mech.* **1961**, *28*, 544–550. [CrossRef]

33. Goodno, B.J.; Gere, J.M. *Mechanics of Materials*, 9th ed.; Thomson Brooks Cole Learning: Boston, MA, USA, 2020.

34. Hertzberg, R.W.; Vinci, R.P.; Hertzberg, J.L. *Deformation and Fracture Mechanics of Engineering Materials*, 6th ed.; John Wiley and Sons Inc.: Hoboken, NJ, USA, 2020.

35. Reddy, J.N. *An Introduction to the Finite Element Method*, 4th ed.; McGraw-Hill: New York, NY, USA, 2020.

36. Courant, R. Variational methods for the solution of problems of equilibrium and vibrations. *Bull. Am. Math. Soc.* **1943**, *49*, 1–23. [CrossRef]

37. Matsuishi, M.; Endo, T. Fatigue of metals subjected to varying stress. *Jpn. Soc. Mech. Eng.* **1968**, *68*, 37–40.

38. Janssens, K.G.F. Universal cycle counting for non-proportional and random fatigue loading. *Int. J. Fatigue* **2020**, *133*, 105409. [CrossRef]

39. Palmgren, A.G. Die Lebensdauer von Kugellagern. *Z Ver Dtsch Ing.* **1924**, *68*, 339–341.

40. IEEE Standard Glossary of Software Engineering Terminology. IEEE STD 610.12-1990. Standards Coordinating Committee of the Computer Society of IEEE. (reaffirmed September 2002). Available online: https://ieeexplore.ieee.org/document/159342 (accessed on 31 December 1990).

41. Klutke, G.; Kiessler, P.C.; Wortman, M.A. A critical look at the bathtub curve. *IEEE Trans. Reliab.* **2015**, *52*, 125–129. [CrossRef]

42. Kreyszig, E. *Advanced Engineering Mathematics*, 10th ed.; John Wiley and Son: Hoboken, NJ, USA, 2011; p. 683.

43. Grove, A. *Physics and Technology of Semiconductor Device*, 1st ed.; Wiley International Edition: New York, NY, USA, 1967; p. 37.

44. Minges, M.L. *Electronic materials handbook*; ASM International: Cleveland, OH, USA, 1989; Volume 1, p. 888.

45. Karnopp, D.C.; Margolis, D.L.; Rosenberg, R.C. *System Dynamics: Modeling, Simulation, and Control of Mechatronic Systems*, 6th ed.; John Wiley & Sons: New York, NY, USA, 2012.

46. Abernethy, R.B. *The New Weibull Handbook*; Reliability Analysis Center: North Palm Beach, FL, USA, 2000; pp. 6–12.

47. Woo, S.; Pecht, M. Failure analysis and redesign of a helix upper dispenser. *Eng Fail Anal.* **2008**, *15*, 642–653. [CrossRef]

48. Woo, S.; O'Neal, D.; Pecht, M. Design of a hinge kit system in a Kimchi refrigerator receiving repetitive stresses. *Eng. Fail. Anal.* **2009**, *16*, 1655–1665. [CrossRef]

49. Woo, S.; O'Neal, D.; Pecht, M. Failure analysis and redesign of the evaporator tubing in a Kimchi refrigerator. *Eng. Fail. Anal.* **2010**, *17*, 369–379. [CrossRef]

50. ASTM A439/A439M–18. *Standard Specification for Austenitic Ductile Iron Castings*; ASTM International: West Conshohocken, PA, USA, 2018.

51. Nadot, Y.; Mendez, J.; Ranganathan, N. Influence of casting defects on the fatigue limit of nodular cast iron. *Int. J. Fatigue* **2004**, *26*, 311–319. [CrossRef]
52. Tang, L.C. Multiple-steps Step-stress Accelerated Life Test. In *Handbook of Reliability Engineering*; Springer: London, UK, 2003; Chapter 24; pp. 441–455.
53. ASTM A874/A874M-98. *Standard Specification for Ferritic Ductile Iron Castings Suitable for Low-Temperature Service*; ASTM International: West Conshohocken, PA, USA, 2018.
54. Neugebauer, G.H. Stress concentration factors and their effect in design. *Prod. Eng. NY A* **1943**, *14*, 82–87.
55. Cinkilic, E.; Yan, X.; Luo, A.A. Modeling Precipitation Hardening and Yield Strength in Cast Al-Si-Mg-Mn Alloys. *Metals* **2020**, *10*, 1356. [CrossRef]
56. Szymczak, T.; Gumienny, G.; Klimek, L.; Goły, M.; Pacyniak, T. Microstructural Characteristics of AlSi9Cu3(Fe) Alloy with High Melting Point Elements. *Metals* **2020**, *10*, 1278. [CrossRef]
57. Li, Q.; Xie, L. Analysis and Optimization of Tooth Surface Contact Stress of Gears with Tooth Profile Deviations, Meshing Errors and Lead Crowning Modifications Based on Finite Element Method and Taguchi Method. *Metals* **2020**, *10*, 1370. [CrossRef]
58. Miao, X.; Liu, X.; Lu, P.; Han, J.; Duan, W.; Wu, M. Influence of Scanning Strategy on the Performances of GO-Reinforced Ti6Al4V Nanocomposites Manufactured by SLM. *Metals* **2020**, *10*, 1379. [CrossRef]
59. Raja, A.; Chukka, S.T.; Jayaganthan, R. Prediction of Fatigue Crack Growth Behavior in Ultrafine Grained Al 2014 Alloy Using Machine Learning. *Metals* **2020**, *10*, 1349. [CrossRef]

Review

Fatigue Crack Growth Analysis with Extended Finite Element for 3D Linear Elastic Material

Yahya Ali Fageehi

Mechanical Engineering Department, Jazan University, P.O. Box 114, Jazan 45142, Saudi Arabia; yfageehi@jazanu.edu.sa

Abstract: This paper presents computational modeling of a crack growth path under mixed-mode loadings in linear elastic materials and investigates the influence of a hole on both fatigue crack propagation and fatigue life when subjected to constant amplitude loading conditions. Though the crack propagation is inevitable, the simulation specified the crack propagation path such that the critical structure domain was not exceeded. ANSYS Mechanical APDL 19.2 was introduced with the aid of a new feature in ANSYS: Smart Crack growth technology. It predicts the propagation direction and subsequent fatigue life for structural components using the extended finite element method (XFEM). The Paris law model was used to evaluate the mixed-mode fatigue life for both a modified four-point bending beam and a cracked plate with three holes under the linear elastic fracture mechanics (LEFM) assumption. Precise estimates of the stress intensity factors (SIFs), the trajectory of crack growth, and the fatigue life by an incremental crack propagation analysis were recorded. The findings of this analysis are confirmed in published works in terms of crack propagation trajectories under mixed-mode loading conditions.

Keywords: XFEM; ANSYS mechanical; smart crack growth; stress intensity factors; LEFM; fatigue life prediction

Citation: Fageehi, Y.A. Fatigue Crack Growth Analysis with Extended Finite Element for 3D Linear Elastic Material. *Metals* **2021**, *11*, 397. https://doi.org/10.3390/met11030397

Academic Editor: Ricardo Branco

Received: 2 February 2021
Accepted: 19 February 2021
Published: 1 March 2021

Publisher's Note: MDPI stays neutral with regard to jurisdictional claims in published maps and institutional affiliations.

1. Introduction

Since the end of the Second World War, the failure of materials under stresses even lower than the yield stresses has gained significant attention [1]. Ensuring the stability of critical structures while establishing their safe working condition is vital. In most industries, the accurate estimation of both crack path and fatigue life are crucial in terms of reliability. In various applications, such as aerospace manufacturing and the aviation industry, experimental studies are necessary for fatigue analysis, but, because of high costs, precise computational methods are required for crack propagation analysis to predict the direction of crack growth and fatigue lifetime in both static and dynamic loading conditions [2]. The failure is related to (a) the presence of flaws such as interfaces and cracks, and (b) the nature of fluctuating loads. Cracks tend to initiate and propagate when subjected to fluctuating loads until a point when the structure does not bear the load that contributes to complete failure. These cracks are considered fatigue cracks and the expected life is one of the major parameters to determine the safety of the structure. This is computed by adding the number of loading cycles required to nucleate the fatigue cracks that lead to failure. The calculation of the growth rate of the cracks is usually based on the relation between the range of the stress intensity factors (SIFs) and the cracks' geometry.

The extended finite element method (XFEM) is an alternate way to predict the SIFs using computational methods. In general, the initiation and propagation of cracks must be associated to the SIFs in a complicated state [3–6]. The extended finite element method proposed by Belytschko and Black in 1999 [7] has been widely used in recent studies. It is based on the standard finite element framework and uses a special displacement feature to allow discontinuities to occur, overcoming the need to re-mesh continuously throughout the crack tip expansion process. To evaluate the SIFs, XFEM was used to perform crack growth analysis without updating the mesh [8]. Extensive work was undertaken to develop efficient models to evaluate the fatigue crack growth (FCG) and fatigue life in order to overcome fatigue failures. There are several proposed experimental models, but they become prohibitive both in terms of cost and time. An effective way to reduce the laboratory work, time, and costs is to incorporate a simulation methodology that involves numerical analysis and use of the ANSYS APDL.19.2 extended finite element method. Many fatigue crack problems identified in the literature to date use different computational approaches in simulating simple and complex two- and three-dimensional geometries [2,9–15].

This work used the ANSYS APDL 19.2 XFEM to precisely predict the mixed-mode stress intensity factors along with the associated fatigue life for a modified four-point bending beam and a cracked plate with three holes. In particular, three methods have been widely used to illustrate the fatigue assessment of materials: the fracture mechanics method developed by Paris and Erdogan [16], the strain-life method independently proposed by Coffin [17], and the stress-life (SN) method proposed by Wöhler [18]. The first approach, by which the crack tip can be described separately by the SIFs, was employed in this study for predicting fatigue life. The second approach is suitable in the lower cycle fatigue range, whereas the third, SN approach estimates the time spent to initiate and grow a crack until the component breaks into parts, which requires stress results from a linear static analysis. The main motivation for this work was to make a significant contribution to the use of ANSYS as an alternative tool for simulating fatigue crack propagation problems during mixed-mode loading and to monitor the trajectory of crack growth in cases of the presence of holes in the geometry.

2. Mixed-Mode Fatigue Life Evaluation Procedure Using ANSYS

ANSYS (version 19.2, Ansys, Inc., Canonsburg, PA, USA), can model three kinds of cracks: arbitrary, semi-elliptical, and pre-meshed. The pre-meshed crack method requires a crack front employed by the Smart Crack growth analysis tool, whereby the stress intensity factor is the criterion of failure. The node sets that were rendered were distributed to the front, top, and bottom of the crack. The latest feature presented in ANSYS is the Smart Crack growth mesh-based tetrahedron, which adds the pre-meshed crack requirement after completion, enabling the selection of the type of crack growth. The sphere of influence process can be used in refining the mesh around the crack tip about the geometric edge that passes through the thickness. The geometric regions to be described are the crack tip, the crack top, and the bottom surfaces of the crack; each of these regions is associated with a node set to be used for analysis. The ANSYS software considered mixed-mode loading where the maximum circumferential stress criterion was implemented. The following are the formulas for the direction angle of crack propagation in ANSYS [19,20]:

$$\theta = \cos^{-1}\left(\frac{3K_{II}^2 + K_I\sqrt{K_I^2 + 8K_{II}^2}}{K_I^2 + 9K_{II}^2}\right) \tag{1}$$

where K_I = max K_I during cyclic loading and K_{II} = max K_{II} during cyclic loading.

In ANSYS Mechanical APDL 19.2, by using XFEM, crack growth simulation was restricted to region II of the typical fatigue crack growth graph, which can be represented as:

$$\frac{da}{dN} = C(\Delta K_{eq})^m \tag{2}$$

From Equation (2), fatigue lifecycles may be predicted for crack increments as:

$$\int_0^{\Delta a} \frac{da}{C(\Delta K_{eq})^m} = \int_0^{\Delta N} dN = \Delta N \tag{3}$$

The equivalent range of the stress intensity factor formula is determined as follows [19]:

$$\Delta K_{eq} = \frac{1}{2} \cos \frac{\theta}{2} [(\Delta K_I (1 + \cos \theta)) - 3\Delta K_{II} \sin \theta] \tag{4}$$

where ΔK_I = the stress intensity factor range in mode I loading and ΔK_{II} = the stress intensity factor range in mode II loading.

Based on numerical analysis, there are many methods formulated for evaluating the stress intensity factors. The interaction integral technique is usually the most accurate method that has the ability of estimating K_I and K_{II} separately. ANSYS proposes two methods for computing SIFs: the displacement extrapolation method (DEM) and the interaction integral method (IIM). The second method was adopted because it is numerically easier to implement and has better precision with fewer mesh requirements. This approach uses the domain integral approach [21] where an auxiliary field is used to separate K_I from K_{II}, as this ability is missing in the domain integral itself. The energy release rate in terms of the mixed-mode stress intensity factors K_I, K_{II}, and K_{III} was proposed as [21,22]:

$$J(s) = \frac{K_I^2 + K_{II}^2}{E^*} + \frac{1+v}{E} K_{III}^2 \quad E^* = \left[\begin{array}{l} \dfrac{E}{(1-v^2)} \text{ Plane strain} \\[2mm] E \ \text{ Plane stress} \end{array} \right. \tag{5}$$

The superimposed state Equation (5) becomes:

$$\begin{aligned} J^S(s) &= \frac{1}{E^*}\left[(K_I + K_I^{aux})^2 + (K_{II} + K_{II}^{aux})^2\right] + \frac{1+v}{E}(K_{III} + K_{III}^{aux})^2 \\ &= J(s) + J^{aux}(s) + I(s) \end{aligned} \tag{6}$$

$$I(s) = \frac{1}{E^*}(2K_I K_I^{aux} + 2K_{II} K_{II}^{aux}) + \frac{1+v}{E}(2K_{III} K_{III}^{aux})$$

Here, superscript (S) denotes the superimposed state; $J(s)$ is the domain integral for the actual state; $J^{aux}(s)$ is the domain integral for the auxiliary state; and $I(s)$ is an integral with interacting actual and auxiliary terms.

By setting $K_I^{aux} = 1$ and $K_{II}^{aux} = K_{III}^{aux} = 0$, Equation (6) yields:

$$K_I = \frac{E^*}{2} I(s) \tag{7}$$

By setting $K_{II}^{aux} = 1$ and $K_I^{aux} = K_{III}^{aux} = 0$, and selecting $K_{III}^{aux} = 1$, $K_I^{aux} = K_{II}^{aux} = 0$ gives the relationships between K_{II} and K_{III}:

$$K_{II} = \frac{E^*}{2} I(s) \tag{8}$$

$$K_{III} = \mu \, I(s) \tag{9}$$

where E and μ are the modulus of elasticity and the modulus of rigidity, respectively.

3. Numerical Results and Discussion

3.1. Modified Four-Point Bending Beam

This case corresponds to a single cracked beam with a hole, loaded in the upper two points and constrained in the lower two points, i.e., a modified four-point bending specimen as shown in Figure 1. This refers to a problem of plane strain that was solved numerically in [23,24]. The geometry is $125 \times 30 \times 10 \text{ mm}^3$ in size, and the hole radius, $R = 5.2$ mm, was located 9.3 mm from the left of the original crack and 14.8 mm above it. This specimen was simulated under fatigue loading with a constant amplitude load ratio, R= 0.1, and the quantity of the applied loads were $P = 100$ N. The initial mesh of this geometry is shown in Figure 2. The material for this specimen was cold-rolled SAE 1020 steel with the following properties as shown in Table 1:

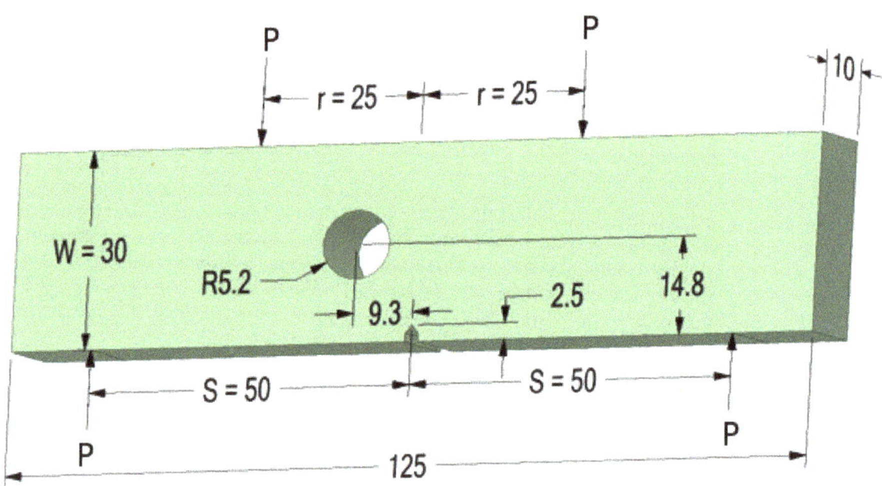

Figure 1. Geometry of the four-point bending beam (dimensions in mm).

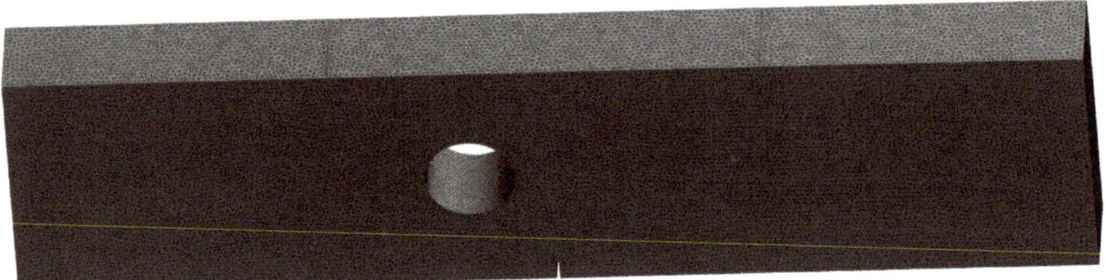

Figure 2. Initial mesh of the four-point bending beam.

Table 1. Materials properties for cold-rolled SAE 1020 steel.

Property	Value in Metric Units
Modulus of elasticity, E	205 GPa
Poisson's ratio, υ	0.333
Yield strength, σ_y	491 MPa
Threshold stress intensity factor, K_{th}	11.6 MPa$\sqrt{m}$
Paris' law coefficient, C	$C = 4.5 \times 10^{-10}$
Paris' law exponent, m	2.1

The predicted crack growth trajectory was smooth and identical to the experimental path predicted by [23] and can be further compared to the predicted trajectories obtained using other numerical methods, such as the finite element method based on local Lepp–Delaunay mesh refinement used in [24], the finite element with configurational forces used in [25], and the coupled extended meshfree–smoothed meshfree method used in [26], as shown in Figure 3a–e, respectively. In the initial period, the crack grew with a small increment when the crack tip was relatively far from the hole. The crack growth direction changed with a large angle and gradually affected the hole with the crack proceeding. Figure 4 illustrates six different steps of the crack growth represented in the von Mises stress distribution, whereas the three-dimensional distribution of the von Mises stress distribution with and without deformation is shown with a legend in Figure 5.

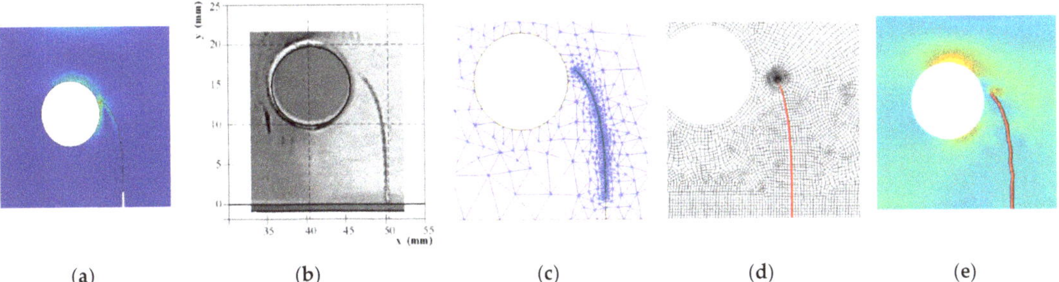

(a)   (b)   (c)   (d)   (e)

Figure 3. Comparison of the crack growth trajectory for the four-point bending beam; (**a**) present study; (**b**) experimental observation reproduced from [23] with permission from Elsevier 2003; (**c**) numerical reproduced from [24] with permission from Elsevier 2010; (**d**) numerical reproduced from [25] with permission from Elsevier 2017; (**e**) numerical reproduced from [26] with permission from Elsevier 2020.

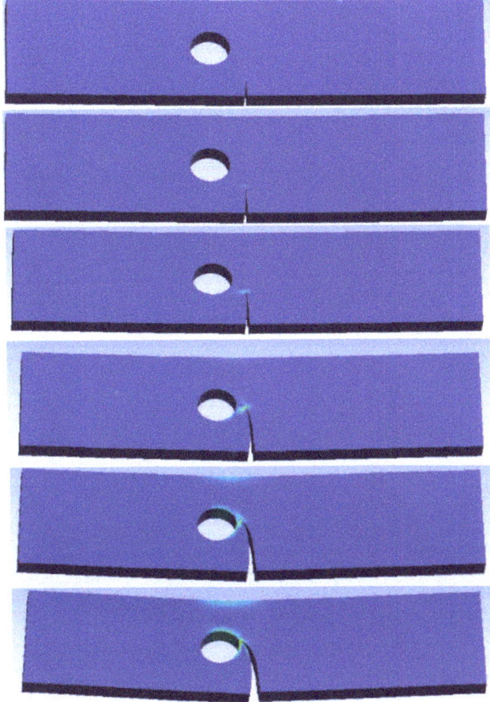

Figure 4. From top to bottom, six different steps of crack growth for the four-point bending beam.

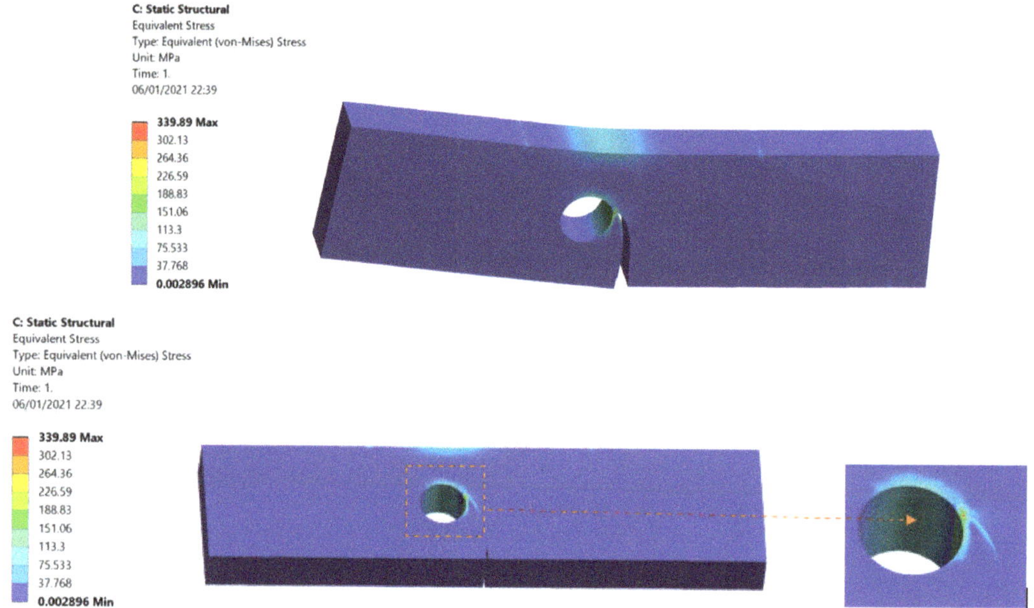

Figure 5. Von Mises stress distribution for the last step of the crack growth with and without deformation.

The distribution of the maximum principal stress is shown in Figure 6 with enlargement of the crack tip area.

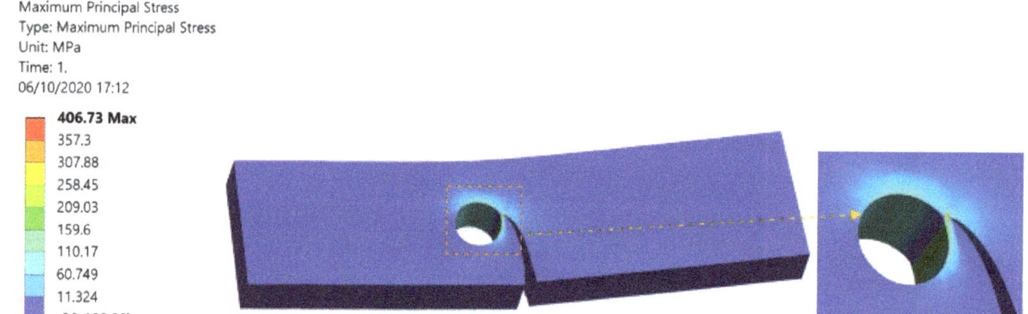

Figure 6. Maximum principal stress distribution.

For fatigue life evaluation, the SIFs are the important criterion. For a normal four-point bending beam, various handbooks may include analytical calculations of the SIFs. For the regular four-point bending beam without a hole the SIFs solution is formulated as follows [27]:

$$K_I = f(a/W)\frac{6P(s-r)\sqrt{\pi a}}{W^2 t} \tag{10}$$

where K_I is the first mode of SIFs, $f(a/W)$ refers to the dimensionless SIF, W is the beam width, t is beam thickness, P is load applied, s and r are the distances defined in Figure 1, and a is the length of the crack. The dimensionless regular stress intensity factor for the point bending beam without holes was formulated as [27]:

$$f(a/W) = \frac{1.1215}{(1-\frac{a}{W})^{(3/2)}}\left[\frac{5}{8} - \frac{5}{12}(a/W) + \frac{1}{8}(a/W)^2 + 5(a/W)^2(1-\frac{a}{W})^6 + \frac{3}{8}\exp(-6.1342(a/W)/(1-\frac{a}{W}))\right] \quad (11)$$

The presence of a hole created a curved crack trajectory in this modified geometry, hence, Equation (11) was no longer valid as a consequence of the curved crack direction. ANSYS can obtain accurate expected $f(a/W)$ values rather than manual solutions for the regular four-point bending beam specimen. In order to achieve the dimensionless stress factor $f(a/W)$, mode I SIFs (K_I) were obtained from ANSYS and substituted into Equation (11). Fitting the fifth-degree polynomial into the stress intensity factors for the modified four-point bending beam gave the following equation:

$$f(a/W) = 12.116(a/W) - 88.937(a/W)^2 + 336.46(a/W)^3 - 595.59(a/W)^4 + 417.66(a/W)^5 + 0.4287 \quad (12)$$

A generalized linear regression method facilitates usage of the formula, which displays SIFs as a function of both the relevant crack and contact parameters, easing assessment of crack growth behavior. For the modified four-point beam specimen used in the above analysis, the numerical dimensionless SIFs were compared with the analytical solution in Equation (11) for the standard beam without a hole, as well as with the dimensionless SIF values calculated by [14] applying the boundary element method (BEM) with BemCracker2D (BC2D) software as shown in Figure 7.

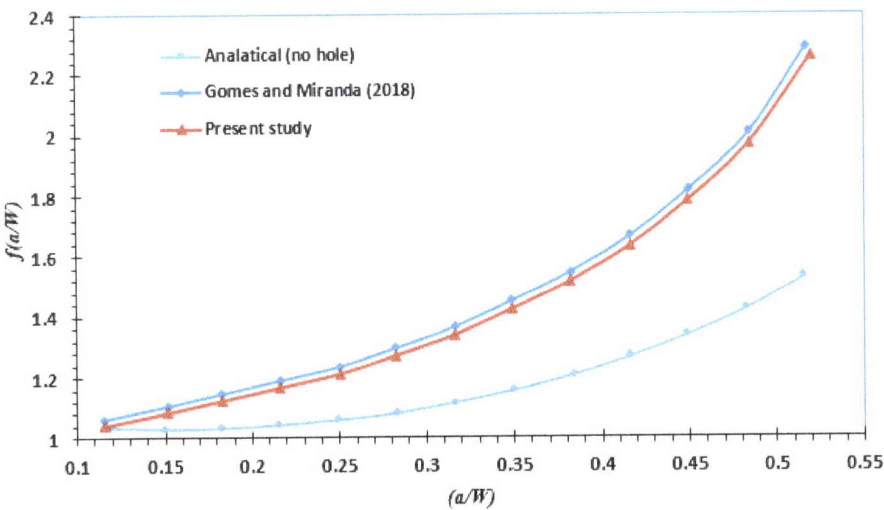

Figure 7. Dimensionless stress intensity factors for the standard and modified four-point bending beams.

The predicted values of both modes of stress intensity factors, i.e., K_I and K_{II} are shown below in Figure 8. As seen in this figure, the crack started to grow in a straight line as the first mode of stress intensity factors dominated the crack growth direction. When the crack direction was influenced by the presence of the hole, the crack grew toward the hole and changed its direction, increasing the values of the second mode of stress intensity factors. The predicted fatigue life according to the number of cycles was compared, as shown in Figure 9, to the experimental results performed by [14] alongside the numerical results for the same authors with two software programs: Vida and BemCracker2D. According to this figure, there was a strong correlation between the present study's result and the Vida software compared to that of the BemCracker2D. According to Figure 8, the bimodality ratio (K_{II}/K_I) was not zero. The direction of the crack was dominated by K_I at the beginning of the crack growth since the K_{II} values were small compared to the K_I values. After that, as the second mode of stress intensity factors, K_{II} was increased gradually up to a maximum value of 21 MPa(mm)$^{1/2}$, leading to a change in the direction of the crack toward the hole.

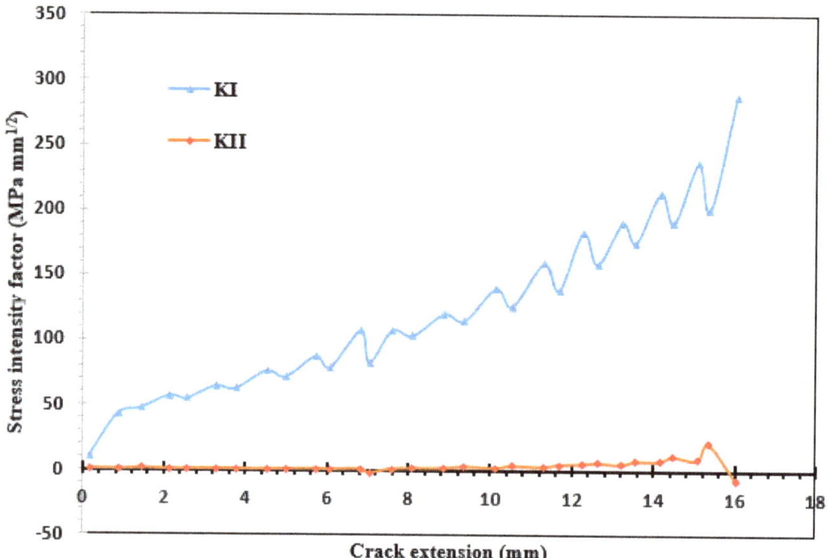

Figure 8. Predicted values of the stress intensity factors.

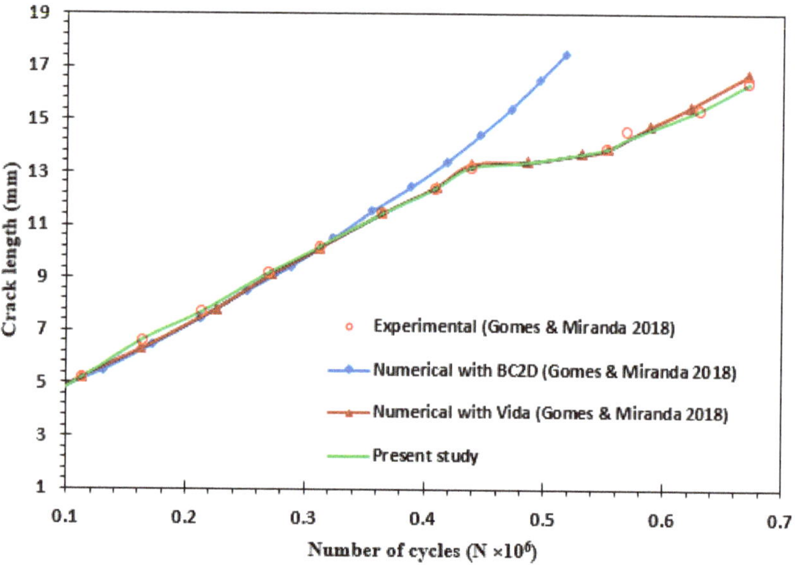

Figure 9. Comparison for fatigue life of the modified four-point bending beam.

3.2. Cracked Plate with Three Holes

Consider a 120 mm × 65 mm × 16 mm rectangular plate with two 13 mm diameter holes near both ends and a 20 mm hole near the middle of the plate, as seen in Figure 10. At the middle of the plate is an initial edge crack of 10 mm. The plate was made from aluminum 7075-T6, with the material properties shown in Table 2, and the amount of the fatigue load was P = 20 kN with a stress ratio R = 0.1. Linear elastic material behavior was assumed. The initial XFEM ANSYS model with an eight-node tetrahedron mesh is shown in Figure 10b. The size of the mesh element was set as 1 mm, creating a mesh of 581,980 nodes and 398,566 elements.

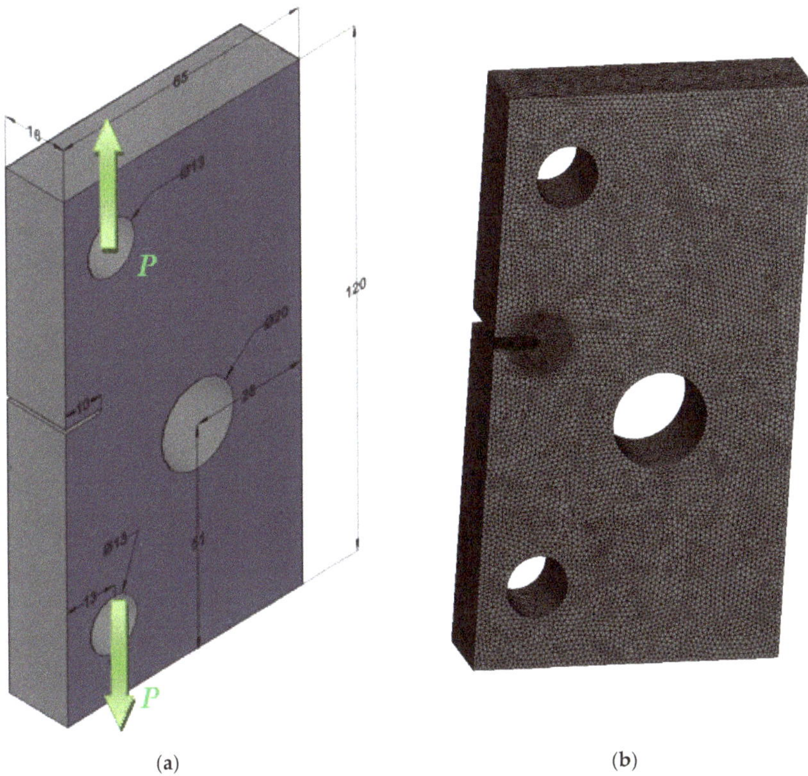

(a)   (b)

Figure 10. (**a**) Description geometry of the cracked plate with three holes (dimensions in mm); (**b**) initial mesh.

Table 2. Materials properties for aluminum 7075-T6.

Property	Value in Metric Units
Modulus of elasticity, E	71.7 GPa
Poisson's ratio, v	0.33
Yield strength, σy	469 MPa
Ultimate strength, σu	538 MPa
Fracture toughness of K_{IC}	938.25 MPa mm$^{0.5}$
Paris' law coefficient, C	5.27×10^{-10}
Paris' law exponent, m	2.947

The crack path growth simulated with ANSYS software was compared, and had strong agreement, with both experimental and numerical results from ABAQUS software obtained by [28] as well as with numerical results performed by [29] using XFEM with a controllable crack propagation strategy, as seen in Figure 11a–d, respectively. The distribution of the maximum principal stress, the von Mises stress, and the equivalent strain are shown in Figures 12–14, respectively.

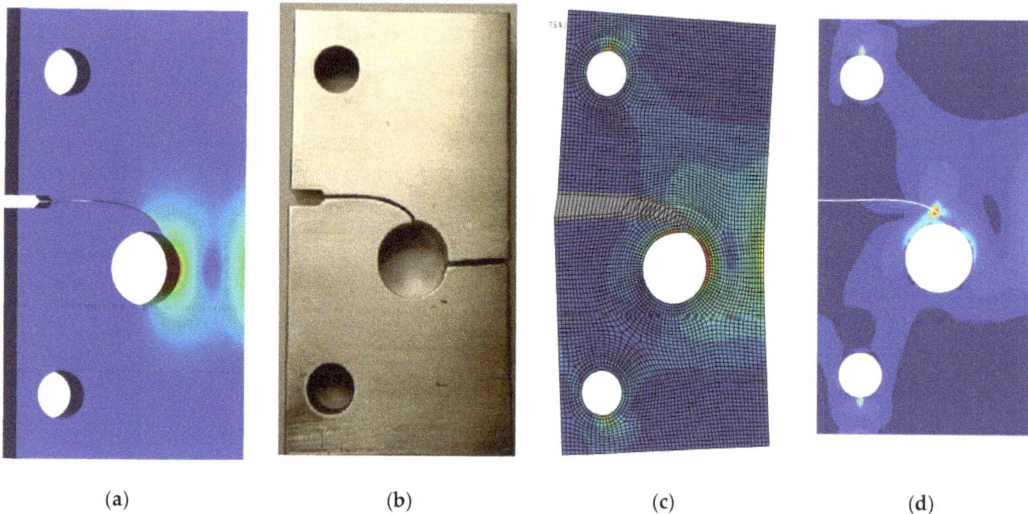

Figure 11. Comparison of crack growth trajectory; (**a**) present study; (**b**) experimental observation reproduced from [28] with permission from Elsevier 2009; (**c**) numerical reproduced from [28] with permission from Elsevier 2009; (**d**) numerical reproduced from [29] with permission from Elsevier 2018.

Figure 12. The maximum principal stress distribution.

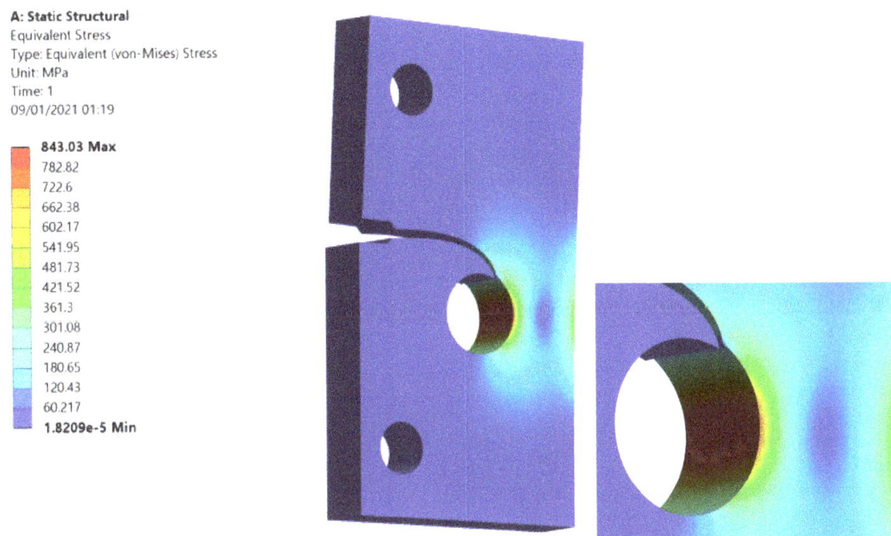

Figure 13. The equivalent von Mises stress distribution.

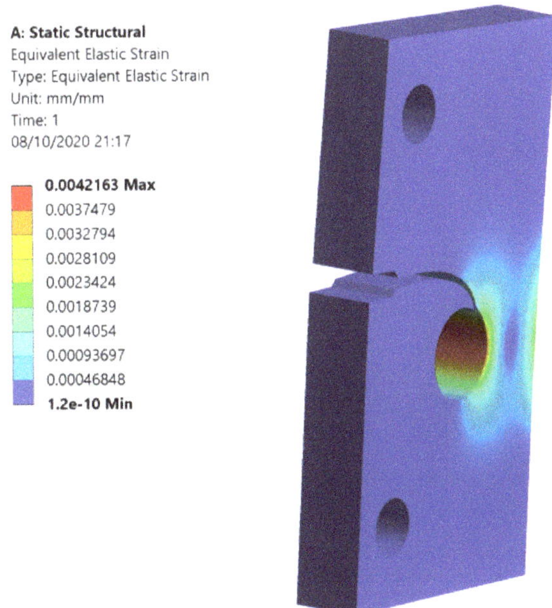

Figure 14. The equivalent elastic strain distribution.

The maximum value of the von Mises stress was in the last step of the crack growth, in which the area around the crack tip is known as a plastic zone. In this area, the behavior of the material is plastic. The plastic zone is created when the stress goes from minimum to maximum values and is called uploading. The plastic zone is plastically elongated in the loading direction. It becomes longer than it was before. As a consequence, the zone is loaded in compression during unloading and reversed plasticity occurs. As pointed out by Rice [30], reversed plasticity requires a local stress increment in the reserved direction in the order of twice the yield stress.

The predicted values of the two modes of stress intensity factors, i.e., K_I and K_{II}, are shown in Figures 15 and 16, respectively. As shown in Figure 14, the crack starts to grow in a straight direction, indicating the domination of K_I followed by a curved direction with an increasing negative value of the second mode, K_{II}, that results in the crack growing toward the hole.

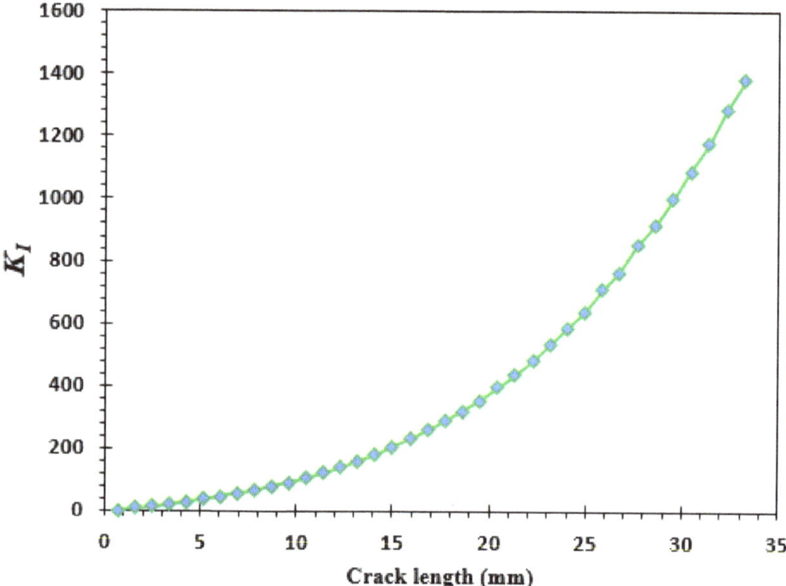

Figure 15. Predicted values of the first mode of stress intensity factors.

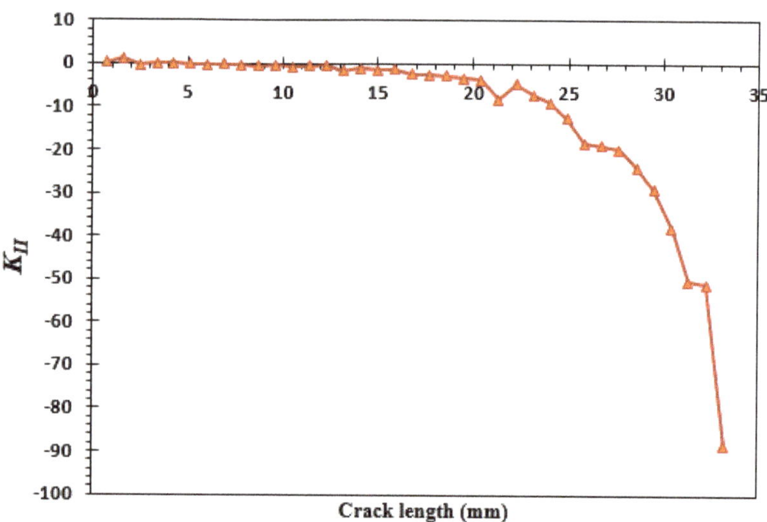

Figure 16. Predicted values of the second mode of stress intensity factors.

4. Conclusions

The crack propagation direction can be specified by simulation techniques such as implementing ANSYS software and ensuring the critical domain of the structure is not exceeded. In order to demonstrate the accuracy and to reveal the merits of the implementation, various numerical examples of crack growth were solved. Different issues were addressed through these examples, such as the effect of hole position on the trajectory of crack growth, mixed-mode stress intensity factors, fatigue lifecycles, and various stress distributions. Depending on the location of the crack, the presence of a hole in the geometry affects the crack and deflects it in the hole's direction, so the crack changes or even passes through the hole and grows until the hole is missing. The validation of the software results were revealed by consistent comparisons with the numerical results of crack propagation by ANSYS and the experimental results.

Funding: This research received no external funding.

Data Availability Statement: Not applicable.

Conflicts of Interest: The author declares no conflict of interest.

References

1. Frost, N.; Pook, L.; Denton, K. A fracture mechanics analysis of fatigue crack growth data for various materials. *Eng. Fract. Mech.* **1971**, *3*, 109–126. [CrossRef]
2. Dirik, H.; Yalçinkaya, T. Crack path and life prediction under mixed mode cyclic variable amplitude loading through XFEM. *Int. J. Fatigue* **2018**, *114*, 34–50. [CrossRef]
3. Sih, G.; Liebowitz, H. *Mathematical Fundamentals in Fracture*; Academic Press: New York, NY, USA, 1968; pp. 67–190.
4. Hellan, K. *Introduction to Fracture Mechanics*; McGraw-Hill: New York, NY, USA, 1985.
5. Barsom, J.; Rolfe, S. *Fracture and Fatigue in Structure: Application of Fracture Mechanics*; American Society for Testing and Materials: Philadelphia, PA, USA, 1999.
6. Broek, D. *Elementary Engineering Fracture Mechanics*; Springer Science & Business Media: Berlin/Heidelberg, Germany, 2012.
7. Belytschko, T.; Black, T. Elastic crack growth in finite elements with minimal remeshing. *Int. J. Numer. Methods Eng.* **1999**, *45*, 601–620. [CrossRef]
8. Bergara, A.; Dorado, J.I.; Martin-Meizoso, A.; Martínez-Esnaola, J.M. Fatigue crack propagation in complex stress fields: Experiments and numerical simulations using the Extended Finite Element Method (XFEM). *Int. J. Fatigue* **2017**, *103*, 112–121. [CrossRef]
9. Alshoaibi, A.M. Finite element procedures for the numerical simulation of fatigue crack propagation under mixed mode loading. *Struct. Eng. Mech.* **2010**, *35*, 283–299. [CrossRef]
10. Alshoaibi, A.M. Comprehensive Comparisons of Two and Three Dimensional Numerical Estimation of Stress Intensity Factors and Crack propagation in Linear Elastic Analysis. *Int. J. Integr. Eng.* **2019**, *11*, 45–52. [CrossRef]
11. Alshoaibi, A.M.; Fageehi, Y.A. 2D finite element simulation of mixed mode fatigue crack propagation for CTS specimen. *J. Mater. Res. Technol.* **2020**, *9*, 7850–7861. [CrossRef]
12. Fageehi, Y.A.; Alshoaibi, A.M. Numerical Simulation of Mixed-Mode Fatigue Crack Growth for Compact Tension Shear Specimen. *Adv. Mater. Sci. Eng.* **2020**. [CrossRef]
13. Chen, H.; Wang, Q.; Zeng, W.; Liu, G.R.; Sun, J.; He, L.; Bui, T.Q. Dynamic brittle crack propagation modeling using singular edge-based smoothed finite element method with local mesh rezoning. *Eur. J. Mech. A/Solids* **2019**, *76*, 208–223. [CrossRef]
14. Gomes, G.; Miranda, A.C. Analysis of crack growth problems using the object-oriented program bemcracker2D. *Frat. Ed Integrità Strutt.* **2018**, *12*, 67–85. [CrossRef]
15. Fageehi, Y.A.; Alshoaibi, A.M. Nonplanar Crack Growth Simulation of Multiple Cracks Using Finite Element Method. *Adv. Mater. Sci. Eng.* **2020**. [CrossRef]
16. Paris, P.; Erdogan, F. A critical analysis of crack propagation laws. *J. Basic Eng.* **1963**, *85*, 528–533. [CrossRef]
17. Coffin, L. Fatigue and Endurance of Metals. In *Cyclic Deformation and Fatigue of Metals (Russian Translation)*; USSR: Moscow, Russia, 1963; pp. 257–272.
18. Wöhler, A. Versuche zur Ermittlung der auf die Eisenbahnwagenachsen einwirkenden Kräfte und die Widerstandsfähigkeit der Wagen-Achsen. *Z. Bauwes.* **1860**, *10*, 583–614.
19. Bjørheim, F. Practical Comparison of Crack Meshing in ANSYS Mechanical APDL 19.2. Master's Thesis, University of Stavanger, Stavanger, Norway, 2019.
20. ANSYS. Academic Research Mechanical, Release 19.2, Help System. In *Coupled Field Analysis Guide, 2020*; ANSYS, Inc.: Canonsburg, PA, USA, 2020.
21. Walters, M.C.; Paulino, G.H.; Dodds, R.H., Jr. Interaction integral procedures for 3-D curved cracks including surface tractions. *Eng. Fract. Mech.* **2005**, *72*, 1635–1663. [CrossRef]

22. Anderson, T.L. *Fracture Mechanics: Fundamentals and Applications*; CRC Press: Cleveland, OH, USA, 2017.
23. Miranda, A.C.O.; Meggiolaro, M.A.; Castro, J.T.P.; Martha, L.F.; Bittencourt, T.N. Fatigue life and crack path predictions in generic 2D structural components. *Eng. Fract. Mech.* **2003**, *70*, 1259–1279. [CrossRef]
24. Azócar, D.; Elgueta, M.; Rivara, M.C. Automatic LEFM crack propagation method based on local Lepp–Delaunay mesh refinement. *Adv. Eng. Softw.* **2010**, *41*, 111–119. [CrossRef]
25. Guo, Y.; Li, Q. Material configurational forces applied to mixed mode crack propagation. *Theor. Appl. Fract. Mech.* **2017**, *89*, 147–157. [CrossRef]
26. Ma, W.; Liu, G.; Wang, W. A coupled extended meshfree–Smoothed meshfree method for crack growth simulation. *Theor. Appl. Fract. Mech.* **2020**, *107*, 102572. [CrossRef]
27. Fett, T. *Stress Intensity Factors, T-Stresses, Weight Functions: Supplement Volume*; KIT Scientific Publishing: Karlsruhe, Germeny, 2009; Volume 55.
28. Giner, E.; Sukumar, N.; Tarancón, J.E.; Fuenmayor, F.J. An Abaqus implementation of the extended finite element method. *Eng. Fract. Mech.* **2009**, *76*, 347–368. [CrossRef]
29. Cheng, Z.; Wang, H. How to control the crack to propagate along the specified path feasibly? *Comput. Methods Appl. Mech. Eng.* **2018**, *336*, 554–577. [CrossRef]
30. Rice, J.R. Mechanics of crack tip deformation and extension by fatigue. In *Fatigue Crack Propagation*; ASTM: West Conshohocken, PA, USA, 1967; pp. 247–309.

MDPI

St. Alban-Anlage 66

4052 Basel

Switzerland

Tel. +41 61 683 77 34

Fax +41 61 302 89 18

www.mdpi.com

Metals Editorial Office

E-mail: metals@mdpi.com

www.mdpi.com/journal/metals